AF595103

Forensic Science and Legal Medicine: A Multidisciplinary Puzzle!

Forensic Science and Legal Medicine: A Multidisciplinary Puzzle!

Editor

Francesco Sessa

MDPI • Basel • Beijing • Wuhan • Barcelona • Belgrade • Manchester • Tokyo • Cluj • Tianjin

Editor
Francesco Sessa
Department of Medical,
Surgical and Advanced
Technologies "G.F. Ingrassia"
University of Catania
Catania
Italy

Editorial Office
MDPI
St. Alban-Anlage 66
4052 Basel, Switzerland

This is a reprint of articles from the Special Issue published online in the open access journal *Healthcare* (ISSN 2227-9032) (available at: www.mdpi.com/journal/healthcare/special_issues/Forensic_Science_Legal_Medicine).

For citation purposes, cite each article independently as indicated on the article page online and as indicated below:

LastName, A.A.; LastName, B.B.; LastName, C.C. Article Title. *Journal Name* **Year**, *Volume Number*, Page Range.

ISBN 978-3-0365-4194-5 (Hbk)
ISBN 978-3-0365-4193-8 (PDF)

Contents

About the Editor

Francesco Sessa

Francesco Sessa is a Lecturer of Forensic Pathology and Legal Medicine at the Department of Medical, Surgical and Advanced, Technologies "G.F. Ingrassia"of the University of Catania. He is a Technical Consultant in the Field of Forensic Genetics for several Italian Courts. He has hold a Ph.D. in Translational Medicine at the University of Foggia, studying the adverse effects linked to the use/abuse of Anabolic-androgenic Steroids (Title of the Ph.D. project: Lifestyle, Medicine, and Anabolic Steroids: Organs Damages, New Molecular Biomarkers, Epidemiology, Pathology, and Toxicology). He is a Member of GE.F.I., the Italian Scientific Society that belonged to the International Society for Forensic Genetics. He is a Member of the American Academy of Forensic Science. He is the author of more than 112 papers published in the indexed journals in Scopus/WOS (H-index 25; Scopus ID=57190022644).

Preface to "Forensic Science and Legal Medicine: A Multidisciplinary Puzzle!"

Even if the terms "forensic sciences"and "legal medicine"seem to be synonymous, they could be defined as two sides of the same coin. It has been demonstrated that they are different components of the application of medical knowledge upon the legal system. Legal medicine has greater relevance to civil and tort law, impacting upon patient care, whereas forensic medicine relates to criminal law and damage to (or by) patients.

The discipline of forensic science is remarkably complex and includes methodologies ranging from DNA analysis to chemical composition and pattern recognition. Many forensic practices were developed under the auspices of law enforcement and vetted primarily by the legal system rather than being subjected to scientific scrutiny and empirical testing. Furthermore, the recent experience of the COVID-19 pandemic suggests a pivotal role of the forensic autopsy in order to gather information about unknown diseases. As previously described, only a full autopsy can investigate the potential mechanisms of damage to organs or systems not readily accessible to biopsies, such as central nervous system or cardiovascular system, leading to appropriate healthcare strategies that are useful in the control of the disease. Another important perspective that underlines the importance of forensic sciences is related to the development of vaccine candidates and new therapies for the prevention and treatment of different diseases, with certain benefits for healthcare.

Furthermore, legal medicine plays a pivotal role in risk management. In healthcare, the term "risk management"refers to all processes employed to detect, monitor, assess, mitigate, and prevent risks in healthcare facilities and safeguard patient safety. Considering the importance of this issue, a further aim of this Special Issue is to assess the role and progress of research and training in the field of risk management.

This Special Issue, entitled "Forensic Science and Legal Medicine: A Multidisciplinary Puzzle!", collected accurate and up-to-date scientific information on all aspects of this theme, publishing original investigations, case series and case reports, and reviews in all forensic and medico-legal branches.

Francesco Sessa
Editor

 healthcare 

Review

Sudden Death in Adults: A Practical Flow Chart for Pathologist Guidance

Francesco Sessa [1,*], Massimiliano Esposito [2], Giovanni Messina [1], Giulio Di Mizio [3], Nunzio Di Nunno [4] and Monica Salerno [2]

[1] Department of Clinical and Experimental Medicine, University of Foggia, 71122 Foggia, Italy; giovanni.messina@unifg.it
[2] Department of Medical, Surgical and Advanced Technologies "G.F. Ingrassia", University of Catania, 95121 Catania, Italy; massimiliano.esposito91@gmail.com (M.E.); monica.salerno@unict.it (M.S.)
[3] Forensic Medicine, Department of Law, Economy and Sociology, Campus "S. Venuta", Magna Graecia University, 88100 Catanzaro, Italy; giulio.dimizio@unicz.it
[4] Department of History, Society and Studies on Humanity, University of Salento, 73100 Lecce, Italy; nunzio.dinunno@icloud.com
* Correspondence: francesco.sessa@unifg.it; Tel.: +39-0881-7369-26 or +39-349-395-0396

Abstract: The medico-legal term "sudden death (SD)" refers to those deaths that are not preceded by significant symptoms. SD in apparently healthy individuals (newborn through to adults) represents a challenge for medical examiners, law enforcement officers, and society as a whole. This review aims to introduce a useful flowchart that should be applied in all cases of SD. Particularly, this flowchart mixes the data obtained through an up-to-date literature review and a revision of the latest version of guidelines for autopsy investigation of sudden cardiac death (SCD) in order to support medico-legal investigation. In light of this review, following the suggested flowchart step-by-step, the forensic pathologist will be able to apply all the indications of the scientific community to real cases. Moreover, it will be possible to answer all questions relative to SD, such as: death may be attributable to cardiac disease or to other causes, the nature of the cardiac disease (defining whether the mechanism was arrhythmic or mechanical), whether the condition causing SD may be inherited (with subsequent genetic counseling), the assumption of toxic or illicit drugs, traumas, and other unnatural causes.

Keywords: sudden death (SD); sudden cardiac death (SCD); autopsy; molecular autopsy; genetics; post-mortem investigation; practical flowchart in SD

Citation: Sessa, F.; Esposito, M.; Messina, G.; Di Mizio, G.; Di Nunno, N.; Salerno, M. Sudden Death in Adults: A Practical Flow Chart for Pathologist Guidance. *Healthcare* **2021**, 9, 870. https://doi.org/10.3390/healthcare9070870

Academic Editor: Mariano Cingolani

Received: 9 June 2021
Accepted: 8 July 2021
Published: 9 July 2021

1. Introduction

The medico-legal term "sudden death (SD)" (also called "sudden and unexpected natural death") refers to those deaths that are not preceded by significant symptoms. The term thus used obviously excludes violent or traumatic deaths. There is no universally accepted definition of sudden death, and time periods ranging from 1 to 48 h have been used in several countries. For example, the World Health Organization (WHO) definition of SD is a "death occurring within 24 h after the onset of the symptoms" [1]; while the Association for European Cardiovascular Pathology has defined SD as "a natural death that occurs within 6 h of the beginning of symptoms in an apparently healthy subject or in one whose disease is not so severe that a fatal outcome would be expected" [2].

SD in apparently healthy individuals (from newborn through to adults) represents a very challenging event for medical examiners, law enforcement officers, and society as a whole [3]. It is particularly difficult to interpret epidemiological data, considering the lack of standardization in death certificate coding and the variability in the definition of SD. Some causes of SD are identifiable through collecting several important pieces of evidence during the external examination, crime scene investigation, and autopsy [4–6]. Moreover, anamnesis and clinical data should be collected in order to identify the exact cause of death [7,8]. Nevertheless, in many cases of SD that present to the medical examiner or coroner, all collected data do not reveal the cause of death [9].

Several SDs are not necessarily "unexpected", and some unexpected deaths are not necessarily "sudden": for this reason, it is extremely important that these autopsies be carried out, and that they are conducted properly. Notably, autopsy findings may have profound effects on the lives and welfare of the family of the deceased, law enforcement agencies, hospital authorities, and private corporations, including insurance companies [10].

In this scenario, this narrative review introduces a useful flowchart that should be applied in all cases of SD. In particular, this flowchart was obtained combining the data obtained through an up-to-date literature review and a revision of the latest version of guidelines for autopsy investigation of sudden cardiac death (SCD) [11] in order to support medico-legal investigations. The choice of the articles for this narrative review was made after evaluation by the authors following a screening of abstracts in the PUBMED database, using the following search terms: "Causes of SD", "SD and Cardiovascular system", "SD and Respiratory System", "SD and Central Nervous System", "SD and Abdominal causes", "SD and Endocrine System", "SD and Iatrogenic", and "SD and Miscellaneous".

2. Causes of Sudden Death

The most prevalent cause of death in the case of SD is related to cardiovascular diseases; nevertheless, when a subject suddenly dies, and the pathologist after the post-mortem examination is not able to identify abnormalities of the cardiac anatomy, a variety of conduction abnormalities without morphological evidence visible at autopsy may be suspected. In light of these considerations, other organs may be involved: these cases of SDs are usually defined as non-cardiac sudden death (nc-SD).

Based on these considerations, SD may be classified under the criteria of the anatomical system involved. It is clear that following these criteria, some degree of overlapping is inevitable. In Figure 1, a system of classification of SD is proposed.

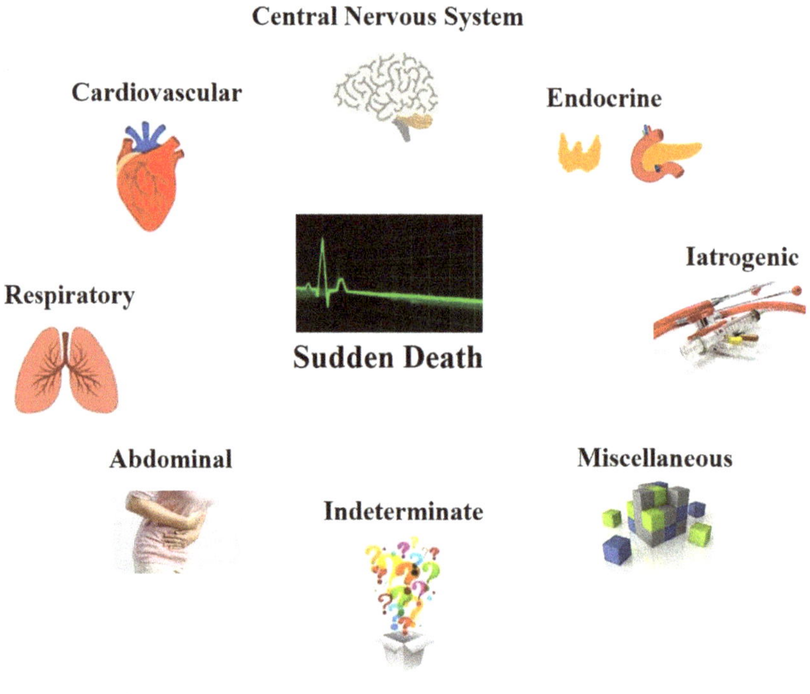

Figure 1. A classification of the possible cause of death in cases of SD.

2.1. Cardiovascular System

The first cause of SD worldwide is cardiovascular diseases, accounting for approximately 90% of such cases [9]. These data are referred to developed countries such as the USA, Japan, and various European countries. When the cardiovascular system is involved, SCD is used. The definitions of SCD are not the same across the scientific community [12]. The incidence of SCD increases with age, varying from about 1 per 1000 per year in subjects of 35–40 years, 2 per 1000 per year by 60 years, and 200 per 1000 per year in the elderly [13,14]. The use of post-mortem imaging is very important in the classification of SCD. Based on these criteria, they may be divided into coronary and non-coronary causes:

- coronary artery diseases (CAD) represent the majority of cardiovascular deaths, even if the data about the percentage are various, ranging from 56.87% [15] to 80% [9]. The percentage of CAD deaths is closely related to age: indeed, it was highest in subjects over 40 years old. CAD may be further divided into atherosclerotic and non-atherosclerotic types. In this subdivision, the atherosclerotic type accounts for most cases, while non-atherosclerotic coronary artery diseases are included in congenital abnormalities, embolism, arteritis, dissecting aneurysms, and external compression or ostial obstruction;
- non-coronary cardiovascular diseases are strictly related to congenital anomalies, valvular heart diseases such as rheumatic heart disease and syphilis, hypertensive heart disease, myocarditis, ruptured aortic aneurysm (acute aortic dissection), and cardiomyopathy [16]. In a simplistic classification, SD due to cardiac genetic alterations could be subdivided into two main groups, channelopathies, and cardiomyopathies [17,18].

The so-called SCDs in the majority of cases are due to coronary artery disease. The post-mortem findings both at the gross examination and at the histological investigations frequently support this kind of diagnosis, describing a clinical picture of severe coronary artery atherosclerosis. They may be associated with coronary artery thrombosis, recent myocardial infarction, and myocardial fibrosis subsequent to different events such as infarction. Nevertheless, these findings are variable and relatively infrequent: for these reasons, they may not be considered decisive to validate the diagnosis [19].

The autopsy finding of critical coronary stenosis (defined as one or more of the major extramural coronary arteries with more than 75% narrowing of the luminal cross-section) is sufficient to invoke a diagnosis of SCD, and this is consistently detected in 90% or more of these patients. Death is believed to be due to rhythm disorders, i.e., dysrhythmias, in most of these cases. In the adult, it is estimated that 10 to 25% of SCD could be related to cardiac channelopathies [20].

Several risk factors for SCD have been reported [21], with age and sex representing two important factors: indeed, the risk of SD is greater in males and obviously increases with age. The death rate improves significantly in middle-old age, especially from age 45 to 64 years [22]. Another important factor is the presence of previous coronary artery disease [23]. Patients with known coronary artery disease had a fourfold greater incidence of SD. Nevertheless, about 55% of those dying suddenly had manifested no prior evidence of coronary artery disease. Another important heart disease closely related to SCD is left ventricular hypertrophy: as previously described, patients with ECG evidence of left ventricular hypertrophy had a 5-fold increased incidence of SD.

In this scenario, hypertension and blood cholesterol may be considered two important indirect risk factors in the insurgence of SCD events. Notably, men with systolic blood pressures >160 mm Hg have an incidence of SD three times greater than those who have systolic pressures <140 mm Hg; moreover, elevated cholesterol levels are generally regarded as a risk factor, even if, to date, no stepwise trend proportional to serum cholesterol has been noted [24]. Obviously, overweight and obese status is related to SCD. It has been described that the risk of this tragic event increases progressively with increased weight, arriving at more than doubled for those weighing 120% or more than their ideal weight [25]. Other important indirect factors that could be considered are cigarette smoking and alcohol/drug abuse. In particular, smokers had a 3-fold greater incidence of SD than non-smokers; moreover, the abuser (meaning smokers of >1 pack per day) had higher rates than did smokers of <1 pack per day [26].

In this context, an infrequent but always tragic event occurs when SD happens in an apparently healthy young adult from spontaneous causes. Although the causes of SD in young subjects are scarce, the most prevalent cause of death is related to cardiovascular disease, with primary arrhythmogenic disorders, atherosclerotic events, cardiomyopathies, and myocarditis as the main contributors. Indeed, in a recent article by Vos et al. [27], cardiovascular diseases and genetic arrhythmias accounted for about 50% of their SD cases. Indeed, based on international data, it is estimated that up to one-third of infantile and juvenile SCD may be explained by cardiac channelopathies [28–31]. The most frequent channelopathies include the long QT syndrome (LQTS), short QT syndrome (SQTS), catecholaminergic polymorphic ventricular tachycardia (CPVT), and Brugada syndrome (BrS) [32–34]. Moreover, in the case of SCD in young people, it could be possible to detect pathogenic mutations in genes encoding structural proteins. These mutations could determine several diseases such as hypertrophic cardiomyopathy, dilated cardiomyopathy, and arrhythmogenic cardiomyopathy (AC) [35]. The majority of these cardiomyopathies may be diagnosed at autopsy, considering that usually, anatomo-morphological changes in cardiac tissue are detected, especially in young adults [36]. Indeed, as recently described, a primary myocardial fibrosis at autopsy is strictly related to variants in genes associated with arrhythmogenic right ventricular cardiomyopathy, dilated cardiomyopathy, and hypertrophic cardiomyopathy; when autopsy does not show these findings, primary myocardial fibrosis may represent an alternative phenotypic expression of structural disease-associated genetic variants, or that risk-associated fibrosis was expressing before the primary disease [37]. Contrariwise, when this tragic event occurs in an infant, a structurally normal heart is usually reported in the post-mortem documentation [38].

One of the most difficult problems for forensic pathologists is the diagnose of SD in subjects with acute cardiac processes that progress rapidly, with non-specific symptoms, leading to death without evident morphological alterations. In these cases, innovative approaches are frequently proposed. For example, post-mortem magnetic resonance imaging could be one of the most promising tools to identify cardiac pathological alterations, highlighting evidence otherwise not visible with routine autopsy [39]. In the same way, the use of biochemical markers in cadaver fluids is frequently investigated as complementary indicators to help to reach valid conclusions about the cause of death. Although the ideal sampling site is debatable, several studies propose either pericardial fluid or peripheral veins as the location for the biological sample. A recent article suggests that cardiac troponin I (cTnI) values in pericardial fluid and the troponin ratio (pericardial fluid/serum ratio) may be helpful in SCD [40]. Finally, immunohistochemical investigation combined with western blot analysis is used to detect morphological changes in myocardial specimens of fatal SD, quantifying the effects of cardiac expression of inflammatory mediators (CD15, IL-1β, IL-6, TNF-α, IL-15, IL-8, MCP-1, ICAM-1, CD18, tryptase) and structural and functional cardiac proteins (troponin I and troponin C) [41].

When SD occurs in a young subject, it may occur during sports activities. In large post-mortem investigations studies of athlete populations in the United States, hypertrophic cardiomyopathy was the most common cardiovascular cause of SD [42]. The second most frequent cardiovascular cause of SD in the subjects practicing sports activities is congenital coronary-artery anomalies: in these cases, the artery originates from the wrong aortic sinus (more commonly, the left main coronary artery originates from Valsalva's right sinus) [43]. Other causes of death are congenital cardiac malformations, such as congenital valvular disease, aortic stenosis, myxomatous mitral valve degeneration (typically associated with Marfan's syndrome), as well as other causes such as myocarditis and coronary atherosclerotic disease [44–46].

A molecular autopsy should be considered fundamental in the case of SD in young people, seeking to always incorporate genetic testing into the post-mortem examination.

Moreover, it has been proposed that community-based data aggregation and sharing should be mandatory, leading to an improved classification of genetic variants [47,48].

2.2. Respiratory System

The most important cause of death in the case of nc-SD involving the respiratory system is acute pulmonary embolism (APE). The symptoms of APE are various with a complex clinical picture. Indeed, APE is characterized by numerous clinical manifestations with a complex interaction between different organs. In this scenario, it is very difficult to make an immediate diagnosis [49].

Fatal pulmonary embolism (PE) represents the common cause of SD related to the respiratory system, usually resulting from a complication of deep venous thrombosis (DVT). Typical symptom and signs of PE is angina with pleuritic chest pain as a consequence of pleural involvement due to pulmonary infarction [50,51]. In fulminant PE, up to 90% of cardiac arrests occur within 1 to 2 h after the onset of symptoms [52]. The mortality rate related to PE is high and it is due to the principal causes: pulmonary mainstream obstruction and liberation of vasoconstrictive mediators from the thrombi [53]. For example, pulmonary thromboembolism has been identified as one of the common clinical pictures of COVID-19, justifying SD in several subjects who died from SARS-CoV-2 infection [54–56]. Other clinical pictures of APE are: symptoms similar to acute respiratory distress syndrome (ARDS) [57]; fever syndrome with or without pseudopneumonia [58]; acute right heart failure/shock/hypotension (often with epigastric pain) [59]; left heart failure (with pulmonary congestion) [60]; chest pain similar to pleuritic syndrome with or without hemoptysis (with or without effusion) [61]; similar to acute coronary syndrome (ACS) (with or without chest pain) [62]; syncope [63]; complete atrioventricular (AV) block with idioventricular rhythm [49]; persistent or paroxysmal atrial fibrillation (AF), atrial flutter, atrial tachycardia [64]; paroxysmal supraventricular tachycardia (PSVT) [49]; DVT and silent PE [65]; platypnea-orthodeoxia [66]; abdominal pain without acute abdomen [64]; and delirium [67].

Other causes of SD strictly related to the respiratory system are:

- Massive hemoptysis. The common causes of massive hemoptysis could be generated by tuberculosis, bronchiectasis, lung abscesses, and mycetomas [68]; another important cause of massive hemoptysis is lung cancer that could generate this symptoms in about 20% of patients [69]. Moreover, cystic fibrosis is reported to be another cause of massive hemoptysis [70].
- Severe pneumonia. This cause is considered one of the natural causes of nc-SD [71]. It could be generated both by viral and bacterial agents, generating different diseases involving myocardial ischemia, a maladaptive response to hypoxia, sepsis-related cardiomyopathy, or other phenomena [72,73]. It is interesting to note that the recent pandemic infection of SARS-CoV-2 could generate severe pneumonia with subsequent cardiac arrest [74].
- Asthma. In a recent study performed in Denmark in young people with uncontrolled asthma, Gullach et al. [75] described that in their cohort, the predominant cause of death was SCD followed by a fatal asthma attack. This clarified the concept that asthma could be considered as a trigger for underlying unknown heart diseases [76].
- Anaphylaxis. Refers to the event cascade that may follow exposure to a particular antigen and causing an acute multi-organ response, with cardiac, coronary/systemic arterial, and dermatological involvement [77]. Cardiovascular symptoms can be the sole manifestation of food allergies, especially in cases where the tragic event occurs during exercise [78]; in similar cases, death may mimic SCD and only a complete autopsy with a full histological and immunohistochemical investigation may disclose the exact cause of death.
- Airway obstruction. Another important cause of nc-SD is hypoxia secondary to pulmonary processes, including small and large airway obstruction (bronchospasm, aspiration, foreign body, edema). Treating the cause of hypoxia/hypoxemia must

be done quickly because this is one of the potentially reversible causes of cardiac arrest [79]. Proper oxygenation and ventilation are key to restoring adequate amounts of oxygen into the system and negating lethal cardiac rhythm [80]. It is important to note that accidental aspiration of an element into the airways is a widespread clinical scenario among children under 3 years old, and it represents the leading cause of infantile death [81].

2.3. *Central Nervous System*

Among the nc-SDs, an important category is the imbalance of the autonomic nervous system control of the cardiovascular system. For these reasons, several neurological conditions, such as stroke, epileptic attacks and brain trauma, drugs, and catecholamine toxicity, may be related to SD [82].

Stroke represents the first cause of SD in this category; it is possible to distinguish three kinds of stroke:

- Intracerebral hemorrhage (ICH) secondary to hypertension or other causes. ICH represents about 10% of strokes. Hypertension is the most important risk factor in order to determine the risk for ICH: for example, Roberts et al. described two cases who died from non-traumatic ICH [83]. Moreover, in a recent study, Verdecchia et al. remarked its value as an independent prognostic marker for SD in the long-term [84].
- Brain infarction secondary to atherosclerosis or embolism. Moreover, a strong positive relationship exists between decreased heart rate variability (HRV) and SD [85]. Brain infarction is implicated in causing diminished HRV and is strictly associated with symptomatic carotid disease [86].
- Subarachnoid hemorrhage (SAH), secondary to ruptured berry aneurysm or other causes. It represents about 5% of stroke, and smoking, high blood pressure, increasing age, and possibly female sex are independent risk factors for SAH [87].

Other than stroke causes include:

- Bacterial meningitis: despite advances in clinical care, it remains a severe disease with a high risk of complications that may lead to SD [88];
- Epilepsy: it is named sudden unexpected death in epilepsy (SUDEP), referring to the sudden and unexpected death of an epileptic patient with no other health issues during normal activity. The exact cause of SUDEP has not been established yet; however, it is assumed to be caused by multiple organ failure [89];
- Brain tumor: even if it represents a rare event, an undiagnosed primary brain tumor may be considered a risk factor for SD. For example, Riezzo et al. described three cases of SD due to glioblastoma [90].

2.4. *Abdominal Causes*

The abdominal region could be involved in the generation of SD. Indeed, even if these causes are a less common cause of SD when compared to other conditions, they are equally important [91,92]. Particularly, the prevalent cause of death involving the abdominal region is massive bleeding into the peritoneal cavity or gastrointestinal tract: it could be linked to different diseases such as duodenal ulcer, gastric ulcer, ulcerative colitis or diverticulitis, malignancy, ruptured ectopic pregnancy, and ruptured viscus for the presence of bowels (e.g., ovarian cysts) [93–95].

Moreover, nc-SD involving the abdominal region could be generated by other diseases such as acute liver failure [96] or acute pancreatitis [97].

2.5. *Endocrine System*

Even if rare, endocrine diseases should be closely related to nc-SD. Notably, when other causes are excluded, this tragic event may be related to different pathological statuses such as adrenal insufficiency (although infrequent, it may occur in individuals treated for other critical conditions where impairment of corticoadrenal function often happen) [98,99], diabetic coma [100], severe hypothyroidism (myxedema) [101], parathyroid crisis [102,103], thymoma [104], etc.

2.6. Iatrogenic

The cases of nc-SD could be related to iatrogenic causes. In particular, even if it is frequently under-evaluated, it could be generated by problems related to prescription drugs [105]. Other causes may be related to the sudden withdrawal of steroids or other drugs. SD represents a complication of other tragic events related to medical errors, such as anesthesia or mismatched blood transfusion [106].

2.7. Miscellaneous

The so-defined miscellaneous category includes drug abuse: it could be related both to the assumption of controlled or uncontrolled substances [107]. In the first case, it is related to the assumption of legal drugs but with errors in the assumption (principally in older people) or voluntary wrong assumption (self-poisoning or for doping purposes), i.e., the use/abuse of anabolic-androgenic steroids (AAS). To date, AASs are frequently used not only to treat both hormonal diseases and other pathologies characterized by muscle loss, but by young people (athletes or individuals) to improve both physical appearance and performance [108,109]. AAS use/abuse is strictly related to the improved risk for SD [110,111].

In the other case, it is related to the assumption of drugs for "recreational" purposes [112]. For example, cocaine in its various forms could be closely related to fatal cardiac arrhythmia, microvascular injury, and acute myocardial ischemia due to coronary vasospasm are the most important causes of cocaine-related SDs. The cardiotoxicity of cocaine is not limited to massive doses of the drug, and underlying heart disease is not a prerequisite for cocaine-related cardiac deaths. Moreover, cocaine is associated with many health complications, including gastrointestinal ischemia/infarction and hemorrhage. For this reason, its assumption may generate SD [113]. Other recreational drugs such as heroin [114], marijuana [115], potent synthetic cannabinoids [116], as well as psychotropic drugs consumed by young people [117] may be related to SDs. Moreover, the proportion of users of recreational drugs was unexpectedly high, even more prevalent than other cardiovascular risk factors. Toxic effects could play an important role as triggers of SD, particularly in young people.

Other miscellaneous nc-SDs could be generated by anaphylaxis or bacteremic shocks, shock from dread, fright or emotion (vagal inhibition), sickle cell crisis, alcoholism, etc. [118,119].

2.8. Indeterminate

This category is reserved for those cases in which the cause of death remains in doubt even after an exhaustive study. Notably, to date, although the progress in the diagnostic fields, several deaths have been classified with the term 'unascertained': this should be reserved for circumstances where the cause and manner of death remain undetermined after autopsy [120].

3. Discussion

The mission of the forensic pathologist is to establish the exact cause of death. As previously described, in the case of SD, it represents a very complex task, although the scientific community has published professional guidelines, book chapters, and many scientific publications on this issue.

In the definition of SD, it is important to clarify the meaning of the term "unexplained": indeed, in common forensic practice, a "death" is defined "unexplained" only after an adequate post-mortem investigation. Obviously, from a forensic point of view, autopsy findings should be carefully evaluated as well as the crime scene investigation with the relative circumstances and the medical history of the victim. Even if the cause of death could remain unanswered after a thorough forensic investigation, from the medical standpoint it represents a critical situation with dangerous clinical implications. Particularly, SD potentially leaves family members at risk. In this scenario, the scientific community should improve efforts to define good practices in the case of SD. In this way, we suggest a practical forensic workflow that should be applied in every case of SD (Figure 2).

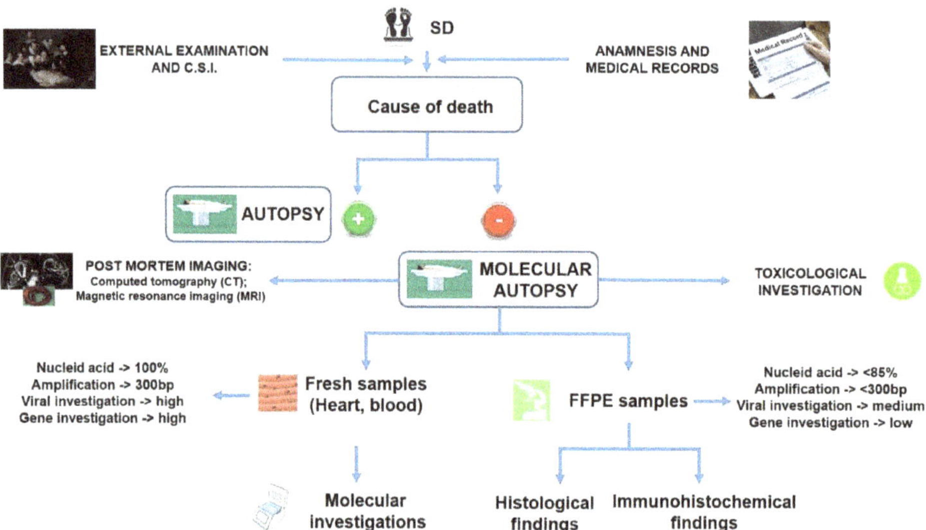

Figure 2. A practical flowchart that should be applied in all cases of SD.

When an SD occurs, several steps should be conducted by the forensic pathologist in order to ascertain the exact cause of death. Obviously, the first two steps that should be performed with greater attention are the external examination and the crime scene investigation (C.S.I.). In the same manner, it is important to collect all information about the anamnesis and/or medical records of the victim. In this regard, several important data should be collected, such as age, gender, occupation, and lifestyle of the victim; the circumstances of death, past medical history, the presence of previous cardiovascular interventions, possible prescription of therapeutical drugs, family cardiac history, possible data collected during the rescue intervention (ECG tracking, serological data, etc.). In all cases, an autopsy should be considered mandatory in order to ascertain the exact cause of death. Before autopsy, the pathologist should collect numerous data from the external examination of the body: it is important to establish body dimensions (weight and height), checking for any dysmorphic features, skin, hair, or skeletal abnormalities, including the presence of a pacemaker or other external interventions (recent intravenous access, intubation, defibrillation, etc.). It is mandatory that all external and internal injuries should be described and photographed. Based on literature data [9], the forensic pathologist is able to directly determine a conclusive cause of death after macroscopic evaluation (positive macroscopic autopsy) in about 65% of cases (green signal), while it remains undefined in the other cases with the so-called "inconclusive autopsy" (negative macroscopic autopsy), indicated with the red signal in the flowchart.

In the cases with a negative macroscopic autopsy, further investigations are needed, applying the so-called molecular autopsy to explain the cause of death. As summarized in this graph, in cases of SD, it should be considered mandatory to perform the so-called "molecular autopsy", meaning the application of molecular techniques to the post-mortem investigation. As summarized in the workflow, the samples collected during a standard autopsy protocol (fresh and/or fixed samples, for example, to perform toxicological and Microscopic Analysis) are used applying molecular techniques to identify hereditary diseases.In this context, it is important to stress the role of post-mortem imaging: several recent publications have reported the pivotal roles of computed tomography (CT) and magnetic resonance imaging (MRI) in order to identify different organ damage before the autopsy [121–123]. The usefulness of traditional X-ray images and post-mortem computed tomography (PMCT) is useful in visualizing calcified plaques, hemopericardium, and valves, and in identifying and locating cardiovascular devices [124,125]. Moreover, by means of a radiological investigation, different heart abnormalities can be highlighted [126]. As recently reported, modern radiological methods, such as multiple detector computed tomography (MDCT), MDCT-angiography, and cardiac MRI have been introduced into post-mortem practice for the investigation of SD, including cases of SCD [127]. A recent retrospective study studied the role of forensic post-mortem CT in order to define the cause of death, especially in cases of acute heart insufficiency or respiratory failure [128].

In the same way, a toxicological investigation plays a pivotal role in excluding the presence of exogenous substances that could be strictly related to SDs. Clinical and forensic toxicology represent two disciplines involving the quantification of xenobiotics in different biological and non-biological samples in order to define the diagnosis, treatment, prognosis and prevention of poisonings and to identify causes or contributory causes of death in cases of fatal intoxications [129]. The samples that could be collected to perform the toxicological investigation are: peripheral venous blood, vitreous humor, hair, urine, bile, pericardial/cerebrospinal fluids, and gastric contents. In a recent cohort study, Bjune et al. [130] reported that SCD victims with positive post-mortem toxicological findings showed polypharmacy assumption, showing this condition may play a proarrhythmic role in these cases. In 2019 Rippoll et al. reported that out of 101 enrolled SD cases, 52 showed positive toxicological findings. Ethanol was the most used substance, followed by legal drugs (meaning therapeutic drugs) and drugs of abuse. In general, the most used toxic substances are illegal drugs (especially cocaine), ethanol, tobacco, doping substances, and therapeutic drugs in not-recommended dosages [131]. Many prescribed drugs or illicit substances exert their adverse effects, both acute and chronic, on heart tissues: for these reasons, toxicological data are strictly related to the histopathological alterations of heart tissue [108,132–135].

As previously described, SD could represent the first manifestation of an unknown inherited cardiac disease. In similar cases, by applying genetic testing, it could be possible to discover the causality, with the identification of family member carriers, adopting preventive strategies [136,137]. Despite the fact that molecular autopsy is recommended in the guidelines for post-mortem investigation of SD, this is rarely performed, maybe because of the necessity to have a specialist laboratory [2,138]. Indeed, it is usually performed in cases of research projects, performing the analysis of the most prevalent genes associated with channelopathies (such as KCNQ1, KCNH2, SCN5A, and others), excluding other important candidate genes [18,139–144]. Moreover, in a recent article by Marey et al. [145], the pivotal role of post-mortem molecular testing in the strategy of family care after SD is remarked, particularly in suspected cardiomyopathy, since genetic findings provide additional useful information for relatives, which are beyond a conventional autopsy. In the last few years, the development of next-generation sequencing (NGS), which represents high-throughput genetic technology, has allowed the investigation of many candidate genes, reducing the bias of untested genomic regions [146]. The main problem related to NGS technologies is the necessity of ultra-specialist personnel and the costs. Indeed, to date, only a few studies have been performed proving the efficacy of this molecular approach in cases of SD [147–149]. In order to perform a molecular investigation, it should be mandatory to sample the victim's body with different specimens, such as blood or other fresh tissues: indeed, the gold standard for molecular genetic testing is EDTA blood or fresh frozen tissue (heart, liver, and spleen) [150]. As summarized in Figure 2, formalin fixation and paraffin embedding (FFPE) of tissues are not recommended for molecular investigations, even if in the last few years technologies have improved the results obtained using these materials. Obviously, these samples are very important in order to perform histological and immunohistochemical investigations. These techniques are commonly used to highlight tissue damage; moreover, in SD cases, they should be used to identify different organ damage and/or the etiology in the myocarditis generated by different external agents, such as viral or bacterial infection: for example, the molecular techniques are the gold standard methods to diagnose viral myocarditis [151–154]. Identification of hereditary diseases is very important to extend the genetic analysis to members of the family of the deceased subject. An extensive, multigenerational family history offers the potential to improve every aspect of care: from establishing a diagnosis, to developing a genetic testing approach, to interpreting genetic test results, to continuously assessing the risk of SCD. Family history reconstruction could be considered not simply a static account of deaths and pre-existing diagnoses, but an ongoing dynamic process that incorporates new and valuable insights from family medical records, clinical cardiology assessments, genetic testing, and visual analysis of the deceased's family tree [155].

In light of these data, following the suggested flowchart step-by-step, the forensic pathologist will be able to apply all the indications of the scientific community to real cases. Particularly, the data discussed in this narrative review suggest that autopsy should be considered mandatory in cases of SD. In this way, it will be possible to answer all questions relative to a SD: such as the death may be attributable to cardiac disease or to other causes, the nature of the cardiac disease (defining whether the mechanism was arrhythmic or mechanical), whether the condition causing SD may be inherited (with the subsequent genetic counseling), the assumption of toxic or illicit drugs, the presence of trauma, and other unnatural causes.

4. Conclusions

Based on this review, it should be stated that a Coroner's post-mortem must be carried out in all cases in which SD occurs if the cause of death is unknown; obviously, in similar cases, the relative medical certificate of cause of death will not be forthcoming. Although forensic investigations may determine the cause of death in most cases, about 19% of cases remain unsolved, requiring further investigation. The molecular autopsy, thanks to modern technologies such as NGS, may help identify the cause of death in a large percentage of unsolved cases. The identification of new risk markers of SD remains one of the most important research fields for the scientific community [85]. For these reasons, a multidisciplinary working group of cardiologists, forensic experts, pathologists, intensive care specialists, geneticists, molecular biologists, and toxicologists is required [156]. Indeed, all clinical and forensic disciplines should build bridges between themselves [129]. Increasing relationships are improving the growth, reliability and the robustness of both kinds of laboratories.

In this context, it is interesting to report the experience of the Swiss Society of Legal Medicine, which in 2015 created a multidisciplinary working group composed of clinical and molecular geneticists together with cardiologists, in the hope of harmonizing the approach to the investigation of SCD [138]. This idea is strongly recommended in the guidelines for autopsy investigation of sudden cardiac death: the authors suggested their adoption throughout Europe with the aim of improving the standards of autopsy practice; moreover, they suggest the development of regional multidisciplinary networks of cardiologists, geneticists, and pathologists to collect more useful information [2]. Molecular medicine represents an important tool to improve the quality of death investigations, providing a new lens to better define the exact cause of death improving traditional methods [157]. Indeed, it is important to stress the key role that the medico-legal investigation has in SD investigations, not only in order to identify the exact cause of death but to indicate prevention in family members: high-resolution variant interpretation provides diagnostic accuracy and healthcare efficiency [158]. The autopsy report should conclude with a clear clinicopathological summary of the major positive findings and their relationship to the cause of death. Indeed, considering the importance of the genetic substrate, particularly in the case of SCD, the identification of mutations of lethal and inheritable cardiomyopathies and cardiac channelopathies can be applied in healthcare management.

Finally, it is important to remark that several causes of death remain unexplained after careful macroscopic, microscopic, and laboratory analyses: in this context, it should be considered mandatory to improve the research activity in this particular field, facilitating the identification of novel causes, and emerging patterns of diseases, causing SCD. In this way, each country should generate a multidisciplinary expert network to allow the interchange of knowledge in order to reduce "unexplained" deaths. Moreover, as recently suggested by Paratz et al. [159], another advisable action that could be adopted is the use of comprehensive multisource surveillance SCD registries that, even while they are not currently widespread, remain an appropriate method.

Author Contributions: Conceptualization, F.S., M.E., G.M., and M.S..; methodology, F.S., G.M., M.E., N.D.N., G.D.M., and M.S.; validation, F.S., G.M., and M.S.; formal analysis, F.S., G.M., M.E., N.D.N., G.D.M., and M.S.; investigation, F.S., G.M., and M.S.; data curation, F.S., G.M., and M.S.; writing—original draft preparation, F.S., G.M., and M.S.; writing—review and editing, F.S., G.M., and M.S. All authors have read and agreed to the published version of the manuscript.

Funding: This research received no external funding.

Institutional Review Board Statement: Not applicable.

Informed Consent Statement: Not applicable.

Data Availability Statement: Data sharing is not applicable to this article.

Acknowledgments: The authors thank the Scientific Bureau of the University of Catania for language support.

Conflicts of Interest: The authors declare no conflict of interest.

References

1. WHO ICD-11 for Mortality and Morbidity Statistics. Available online: https://icd.who.int/browse11/l-m/en#/http%3A%2F%2Fid.who.int%2Ficd%2Fentity%2F426429380 (accessed on 30 June 2021).
2. Basso, C.; Association for European Cardiovascular Pathology; Aguilera, B.; Banner, J.; Cohle, S.; D'Amati, G.; De Gouveia, R.H.; Di Gioia, C.; Fabre, A.; Gallagher, P.J.; et al. Guidelines for autopsy investigation of sudden cardiac death: 2017 update from the Association for European Cardiovascular Pathology. *Virchows Arch.* **2017**, *471*, 691–705. [CrossRef]
3. Tang, Y.; Siegel, D.; Sampson, B. Molecular Investigations of Sudden Unexplained Deaths. *Acad. Forensic Pathol.* **2011**, *1*, 194–201. [CrossRef]
4. Lambert, A.B.E.; Parks, S.E.; Camperlengo, L.; Cottengim, C.; Anderson, R.L.; Covington, T.M.; Shapiro-Mendoza, C.K. Death Scene Investigation and Autopsy Practices in Sudden Unexpected Infant Deaths. *J. Pediatr.* **2016**, *174*, 84–90.e1. [CrossRef] [PubMed]
5. Zhuo, L.; Zhang, Y.; Zielke, H.R.; Levine, B.; Zhang, X.; Chang, L.; Fowler, D.; Li, L. Sudden unexpected death in epilepsy: Evaluation of forensic autopsy cases. *Forensic Sci. Int.* **2012**, *223*, 171–175. [CrossRef] [PubMed]
6. Bennett, T.; Martin, L.J.; Heathfield, L.J. A retrospective study of death scene investigation practices for sudden unexpected death of infants (SUDI) in Cape Town, South Africa. *Forensic Sci. Med. Pathol.* **2019**, *16*, 49–56. [CrossRef] [PubMed]
7. Brugada, R.; Brugada, J.; Brugada, P. *Clinical Approach to Sudden Cardiac Death Syndromes*; Springer: London, UK, 2010; ISBN 9781848829268.
8. Roncati, L.; Manenti, A.; Barbolini, G. Sudden death. In *Advances in Health and Disease*; Nova Science: Hauppauge, NY, USA, 2018; ISBN 9781536130218.
9. Sanchez, O.; Campuzano, O.; Fernández-Falgueras, A.; Sarquella-Brugada, G.; Cesar, S.; Mademont, I.; Mates, J.; Pérez-Serra, A.; Coll, M.; Pico, F.; et al. Natural and undetermined sudden death: Value of post-mortem genetic investigation. *PLoS ONE* **2016**, *11*, e0167358. [CrossRef] [PubMed]
10. Maron, B.J.; Mackey-Bojack, S.; Facile, E.; Duncanson, E.; Rowin, E.J.; Maron, M.S. Hypertrophic Cardiomyopathy and Sudden Death Initially Identified at Autopsy. *Am. J. Cardiol.* **2020**, *127*, 139–141. [CrossRef]
11. Wong, L.C.H.; Behr, E.R. Sudden unexplained death in infants and children: The role of undiagnosed inherited cardiac conditions. *EP Eur.* **2014**, *16*, 1706–1713. [CrossRef] [PubMed]
12. Kong, M.H.; Fonarow, G.; Peterson, E.D.; Curtis, A.B.; Hernandez, A.F.; Sanders, G.D.; Thomas, K.L.; Hayes, D.L.; Al-Khatib, S.M. Systematic Review of the Incidence of Sudden Cardiac Death in the United States. *J. Am. Coll. Cardiol.* **2011**, *57*, 794–801. [CrossRef]

13. Gräsner, J.-T.; Bossaert, L. Epidemiology and management of cardiac arrest: What registries are revealing. *Best Pract. Res. Clin. Anaesthesiol.* **2013**, *27*, 293–306. [CrossRef]
14. Al-Khatib, S.M.; Stevenson, W.G.; Ackerman, M.J.; Bryant, W.J.; Callans, D.J.; Curtis, A.B.; Deal, B.J.; Dickfeld, T.; Field, M.E.; Fonarow, G.C.; et al. 2017 AHA/ACC/HRS Guideline for Management of Patients with Ventricular Arrhythmias and the Prevention of Sudden Cardiac Death: Executive Summary: A Report of the American College of Cardiology/American Heart Association Task Force on Clinical Practice Gui. *J. Am. Coll. Cardiol.* **2018**, *15*, e190–e252. [CrossRef] [PubMed]
15. Bagnall, R.; Weintraub, R.G.; Ingles, J.; Duflou, J.; Yeates, L.; Lam, L.; Davis, A.M.; Thompson, T.; Connell, V.; Wallace, J.; et al. A Prospective Study of Sudden Cardiac Death among Children and Young Adults. *N. Engl. J. Med.* **2016**, *374*, 2441–2452. [CrossRef] [PubMed]
16. Fineschi, V.; Michalodimitrakis, M.; D'Errico, S.; Neri, M.; Pomara, C.; Riezzo, I.; Turillazzi, E. Insight into stress-induced cardiomyopathy and sudden cardiac death due to stress. A forensic cardio-pathologist point of view. *Forensic Sci. Int.* **2010**, *194*, 1–8. [CrossRef] [PubMed]
17. Rodríguez-Calvo, M.S.; Brion, M.; Allegue, C.; Concheiro, L.; Carracedo, A. Molecular genetics of sudden cardiac death. *Forensic Sci. Int.* **2008**, *182*, 1–12. [CrossRef] [PubMed]
18. Pellegrino, P.L.; Bafunno, V.; Ieva, R.; Brunetti, N.D.; Mavilio, G.; Sessa, F.; Grimaldi, M.; Margaglione, M.; Di Biase, M. A Novel Mutation in Human Ether-a-Go-Go-Related Gene, Alanine to Proline at Position 490, Found in a Large Family with Autosomal Dominant Long QT Syndrome. *Am. J. Cardiol.* **2007**, *99*, 1737–1740. [CrossRef] [PubMed]
19. Barletta, V.; Fabiani, I.; Lorenzo, C.; Nicastro, I.; Bello, V. Di Sudden cardiac death: A review focused on cardiovascular imaging. *J. Cardiovasc. Echogr.* **2014**, *24*, 41–51. [CrossRef]
20. Campuzano, O.; Allegue, C.; Partemi, S.; Iglesias, A.; Oliva, A.; Brugada, R. Negative autopsy and sudden cardiac death. *Int. J. Leg. Med.* **2014**, *128*, 599–606. [CrossRef] [PubMed]
21. Bogle, B.M.; Ning, H.; Mehrotra, S.; Goldberger, J.J.; Lloyd-Jones, D.M. Lifetime Risk for Sudden Cardiac Death in the Community. *J. Am. Hear. Assoc.* **2016**, *5*, e002398. [CrossRef] [PubMed]
22. Elkilany, G.; Singh, R.B.; Adeghate, E.; Singh, J.; Bidasee, K.; Shehab, O.; Hristova, K. Sudden cardiac death. *World Heart J.* **2017**, *9*, 51–62.
23. Osman, J.; Tan, S.C.; Lee, P.Y.; Low, T.Y.; Jamal, R. Sudden Cardiac Death (SCD)—Risk stratification and prediction with molecular biomarkers. *J. Biomed. Sci.* **2019**, *26*, 1–12. [CrossRef]
24. Ettehad, D.; Emdin, C.A.; Kiran, A.; Anderson, S.G.; Callender, T.; Emberson, J.; Chalmers, J.; Rodgers, A.; Rahimi, K. Blood pressure lowering for prevention of cardiovascular disease and death: A systematic review and meta-analysis. *Lancet* **2016**, *387*, 957–967. [CrossRef]
25. Hess, P.L.; Al-Khalidi, H.R.; Friedman, D.J.; Mulder, H.; Kucharska-Newton, A.; Rosamond, W.R.; Lopes, R.D.; Gersh, B.J.; Mark, D.B.; Curtis, L.H.; et al. The Metabolic Syndrome and Risk of Sudden Cardiac Death: The Atherosclerosis Risk in Communities Study. *J. Am. Heart Assoc.* **2017**, *6*. [CrossRef]
26. Mukamal, K.J. The Effects of Smoking and Drinking on Cardiovascular Disease and Risk Factors. *Alcohol Res. Health* **2006**, *29*, 199–202.
27. Vos, A.; van der Wal, A.C.; Teeuw, A.H.; Bras, J.; Vink, A.; Nikkels, P.G.J. Cardiovascular causes of sudden unexpected death in children and adolescents (0–17 years): A nationwide autopsy study in the Netherlands. *Netherlands Hear. J.* **2018**, *26*, 500–505. [CrossRef] [PubMed]
28. Tester, D.J.; Ackerman, M.J. The molecular autopsy: Should the evaluation continue after the funeral? *Pediatric Cardiol.* **2012**, *33*, 461–470. [CrossRef] [PubMed]
29. Ackerman, M.J. State of Postmortem Genetic Testing Known as the Cardiac Channel Molecular Autopsy in the Forensic Evaluation of Unexplained Sudden Cardiac Death in the Young. *Pacing Clin. Electrophysiol.* **2009**, *32*, S86–S89. [CrossRef] [PubMed]
30. Semsarian, C.; Ingles, J.; Wilde, A.A. Sudden cardiac death in the young: The molecular autopsy and a practical approach to surviving relatives. *Eur. Heart J.* **2015**, *36*, 1290–1296. [CrossRef]
31. Neri, M.; Di Donato, S.; Maglietta, R.; Pomara, C.; Riezzo, I.; Turillazzi, E.; Fineschi, V. Sudden death as presenting symptom caused by cardiac primary multicentric left ventricle rhabdomyoma, in an 11-month-old baby. An immunohistochemical study. *Diagn. Pathol.* **2012**, *7*, 169. [CrossRef]
32. Magi, S.; Lariccia, V.; Maiolino, M.; Amoroso, S.; Gratteri, S. Sudden cardiac death: Focus on the genetics of channelopathies and cardiomyopathies. *J. Biomed. Sci.* **2017**, *24*, 1–18. [CrossRef]
33. Campuzano, O.; Sarquella-Brugada, G.; Brugada, R.; Brugada, J. Genetics of channelopathies associated with sudden cardiac death. *Glob. Cardiol. Sci. Pract.* **2015**, *2015*, 39. [CrossRef]
34. Van Niekerk, C.; Van Deventer, B.S.; Du Toit-Prinsloo, L. Long QT syndrome and sudden unexpected infant death. *J. Clin. Pathol.* **2017**, *70*, 808–813. [CrossRef]
35. Care, M.; Chauhan, V.; Spears, D. Genetic Testing in Inherited Heart Diseases: Practical Considerations for Clinicians. *Curr. Cardiol. Rep.* **2017**, *19*, 88. [CrossRef]
36. Bezzina, C.R.; Lahrouchi, N.; Priori, S.G. Genetics of Sudden Cardiac Death. *Circ. Res.* **2015**, *116*, 1919–1936. [CrossRef]
37. Junttila, M.J.; Holmström, L.; Pylkäs, K.; Mantere, T.; Kaikkonen, K.; Porvari, K.; Kortelainen, M.L.; Pakanen, L.; Kerkelä, R.; Myerburg, R.J.; et al. Primary Myocardial Fibrosis as an Alternative Phenotype Pathway of Inherited Cardiac Structural Disorders. *Circulation* **2018**, *137*, 2716–2726. [CrossRef]

38. Sarquella-Brugada, G.; Campuzano, O.; Cesar, S.; Iglesias, A.; Fernandez, A.; Brugada, J.; Brugada, R. Sudden infant death syndrome caused by cardiac arrhythmias: Only a matter of genes encoding ion channels? *Int. J. Legal Med.* **2016**, *130*, 415–420. [CrossRef] [PubMed]
39. Cafarelli, F.P.; Macarini, L.; Cipolloni, L.; Maglietta, F.; Guglielmi, G.; Sessa, F.; Pennisi, A.; Cantatore, S.; Bertozzi, G. Ex situ heart magnetic resonance imaging and angiography: Feasibility study for forensic purposes. *Forensic Imaging* **2021**, *25*, 200442. [CrossRef]
40. Hernández-Romero, D.; Valverde-Vázquez, M.D.R.; Hernández Del Rincón, J.P.; Noguera-Velasco, J.A.; Pérez-Cárceles, M.D.; Osuna, E. Diagnostic Application of Postmortem Cardiac Troponin I Pericardial Fluid/Serum Ratio in Sudden Cardiac Death. *Diagnostics* **2021**, *11*, 614. [CrossRef] [PubMed]
41. Turillazzi, E.; Di Paolo, M.; Neri, M.; Riezzo, I.; Fineschi, V. A theoretical timeline for myocardial infarction: Immunohistochemical evaluation and western blot quantification for Interleukin-15 and Monocyte chemotactic protein-1 as very early markers. *J. Transl. Med.* **2014**, *12*, 188. [CrossRef] [PubMed]
42. Ullal, A.J.; Abdelfattah, R.S.; Ashley, E.A.; Froelicher, V.F. Hypertrophic Cardiomyopathy as a Cause of Sudden Cardiac Death in the Young: A Meta-Analysis. *Am. J. Med.* **2016**, *129*, 486–496.e2. [CrossRef]
43. Basso, C.; Maron, B.J.; Corrado, D.; Thiene, G. Clinical profile of congenital coronary artery anomalies with origin from the wrong aortic sinus leading to sudden death in young competitive athletes. *J. Am. Coll. Cardiol.* **2000**, *35*, 1493–1501. [CrossRef]
44. Glorioso, J.; Reeves, M. Marfan syndrome: Screening for sudden death in athletes. *Curr. Sports Med. Rep.* **2002**, *1*, 67–74. [CrossRef]
45. De Backer, J. Cardiovascular characteristics in Marfan syndrome and their relation to the genotype. *Verh. K. Acad. Geneeskd. Belg.* **2009**, *71*, 335–371. [PubMed]
46. Aalberts, J.J.; van Tintelen, J.P.; Meijboom, L.J.; Polko, A.; Jongbloed, J.D.; van der Wal, H.; Pals, G.; Osinga, J.; Timmermans, J.; De Backer, J.; et al. Relation between genotype and left-ventricular dilatation in patients with Marfan syndrome. *Gene* **2014**, *534*, 40–43. [CrossRef] [PubMed]
47. Rueda, M.; Wagner, J.L.; Phillips, T.C.; Topol, S.E.; Muse, E.D.; Lucas, J.R.; Wagner, G.N.; Topol, E.J.; Torkamani, A. Molecular Autopsy for Sudden Death in the Young: Is Data Aggregation the Key? *Front. Cardiovasc. Med.* **2017**, *4*, 72. [CrossRef] [PubMed]
48. Suktitipat, B.; Sathirareuangchai, S.; Roothumnong, E.; Thongnoppakhun, W.; Wangkiratikant, P.; Vorasan, N.; Krittayaphong, R.; Pithukpakorn, M.; Boonyapisit, W. Molecular investigation by whole exome sequencing revealed a high proportion of pathogenic variants among Thai victims of sudden unexpected death syndrome. *PLoS ONE* **2017**, *12*, e0180056. [CrossRef] [PubMed]
49. Morrone, D.; Morone, V. Acute pulmonary embolism: Focus on the clinical picture. *Korean Circ. J.* **2018**, *48*, 365–381. [CrossRef] [PubMed]
50. Stein, P.D.; Henry, J.W. Clinical Characteristics of Patients with Acute Pulmonary Embolism Stratified According to Their Presenting Syndromes. *Chest* **1997**, *112*, 974–979. [CrossRef]
51. Islam, M.; Filopei, J.; Frank, M.; Ramesh, N.; Verzosa, S.; Ehrlich, M.; Bondarsky, E.; Miller, A.; Steiger, D. Pulmonary infarction secondary to pulmonary embolism: An evolving paradigm. *Respirology* **2018**, *23*, 866–872. [CrossRef] [PubMed]
52. Dalen, J.E.; Alpert, J.S. Natural history of pulmonary embolism. *Prog. Cardiovasc. Dis.* **1975**, *17*, 259–270. [CrossRef]
53. Karia, S.; Screaton, N. Pulmonary embolism. In *Imaging*; Barr, R.G., Parr, D.G., Vogel-Claussen, J., Eds.; ERS Monograph: Lausanne, Switzerland, 2015.
54. D'Errico, S.; Zanon, M.; Montanaro, M.; Radaelli, D.; Sessa, F.; Di Mizio, G.; Montana, A.; Corrao, S.; Salerno, M.; Pomara, C. More than Pneumonia: Distinctive Features of SARS-CoV-2 Infection. From Autopsy Findings to Clinical Implications: A Systematic Review. *Microorganisms* **2020**, *8*, 1642. [CrossRef]
55. Cipolloni, L.; Sessa, F.; Bertozzi, G.; Baldari, B.; Cantatore, S.; Testi, R.; D'Errico, S.; Di Mizio, G.; Asmundo, A.; Castorina, S.; et al. Preliminary Post-Mortem COVID-19 Evidence of Endothelial Injury and Factor VIII Hyperexpression. *Diagnostics* **2020**, *10*, 575. [CrossRef] [PubMed]
56. Sessa, F.; Salerno, M.; Pomara, C. Autopsy Tool in Unknown Diseases: The Experience with Coronaviruses (SARS-CoV, MERS-CoV, SARS-CoV-2). *Medicina* **2021**, *57*, 309. [CrossRef] [PubMed]
57. Geibel, A.; Zehender, M.; Kasper, W.; Olschewski, M.; Klima, C.; Konstantinides, S. Prognostic value of the ECG on admission in patients with acute major pulmonary embolism. *Eur. Respir. J.* **2005**, *25*, 843–848. [CrossRef]
58. Xu, K.; Tang, X.; Song, Y.; Chen, Z. The Diagnostic Dilemma between Pulmonary Embolism with Positive Chest Imaging and Pneumonia: A Case Report and Literature Review. *J. Transl. Med. Epidemiol.* **2015**, *3*, 1039.
59. Wood, K.E. Major pulmonary embolism: Review of a pathophysiologic approach to the golden hour of hemodynamically significant pulmonary embolism. *Chest* **2002**, *121*, 877–905. [CrossRef]
60. Manne, J.R.R. Acute ST segment elevation in a patient with massive pulmonary embolism mimicking acute left main coronary artery obstruction. *J. Am. Coll. Cardiol.* **2016**, *67*, 1118. [CrossRef]
61. Kosuge, M.; Ebina, T.; Hibi, K.; Tsukahara, K.; Iwahashi, N.; Gohbara, M.; Matsuzawa, Y.; Okada, K.; Morita, S.; Umemura, S.; et al. Differences in negative T waves among acute coronary syndrome, acute pulmonary embolism, and Takotsubo cardiomyopathy. *Eur. Heart J. Acute Cardiovasc. Care* **2012**, *1*, 349–357. [CrossRef] [PubMed]
62. Omar, H.R. ST-segment elevation in V1-V4 in acute pulmonary embolism: A case presentation and review of literature. *Eur. Heart J. Acute Cardiovasc. Care* **2016**, *5*, 579–586. [CrossRef] [PubMed]
63. Duplyakov, D.; Kurakina, E.; Pavlova, T.; Khokhlunov, S.; Surkova, E. Value of syncope in patients with high-to-intermediate risk pulmonary artery embolism. *Eur. Heart J. Acute Cardiovasc. Care* **2015**, *4*, 353–358. [CrossRef]

64. Inouye, S.K. Delirium in Older Persons. *N. Engl. J. Med.* **2006**, *318*, 1161–1174. [CrossRef]
65. Laack, T.A.; Goyal, D.G. Pulmonary embolism: An unsuspected killer. *Emerg. Med. Clin. N. Am.* **2004**, *22*, 961–983. [CrossRef] [PubMed]
66. Søgaard, K.; Schmidt, M.; Pedersen, L. Thirty-Year Mortality After Venous Thromboembolism: A Population-Based Cohort Study. *J. Vasc. Surg.* **2015**, *61*, 281. [CrossRef]
67. Carrascosa, M.F.; Batán, A.M.; Novo, M.F.A. Delirium and pulmonary embolism in the elderly. *Mayo Clin. Proc.* **2009**, *84*, 91–92. [CrossRef]
68. Corey, R.; Hla, K.M. Major and Massive Hemoptysis: Reassessment of Conservative Management. *Am. J. Med. Sci.* **1987**, *294*, 301–309. [CrossRef] [PubMed]
69. Kvale, P.A.; Selecky, P.A.; Prakash, U.B.S. Palliative care in lung cancer: ACCP evidence-based clinical practice guidelines (2nd edition). *Chest* **2007**, *132*, 368S–403S. [CrossRef] [PubMed]
70. Radchenko, C.; Alraiyes, A.H.; Shojaee, S. A systematic approach to the management of massive hemoptysis. *J. Thorac. Dis.* **2017**, *9*, S1069–S1086. [CrossRef]
71. Fineschi, V.; Turillazzi, E. Natural Causes of Sudden Death: Noncardiac. In *Wiley Encyclopedia of Forensic Science*; Wiley: Hoboken, NJ, USA, 2009. [CrossRef]
72. Smith, M.E.B.; Chiovaro, J.C.; O'Neil, M.; Kansagara, D.; Quiñones, A.R.; Freeman, M.; Motu'apuaka, M.L.; Slatore, C.G. Early warning system scores for clinical deterioration in hospitalized patients: A systematic review. *Ann. Am. Thorac. Soc.* **2014**, *11*, 1454–1465. [CrossRef]
73. Carr, G.E.; Yuen, T.C.; McConville, J.F.; Kress, J.P.; VandenHoek, T.L.; Hall, J.B.; Edelson, D.P. Early Cardiac Arrest in Patients Hospitalized with Pneumonia: A Report From the American Heart Association's Get With the Guidelines-Resuscitation Program. *Chest* **2012**, *141*, 1528–1536. [CrossRef]
74. Shirazi, S.; Mami, S.; Mohtadi, N.; Ghaysouri, A.; Tavan, H.; Nazari, A.; Kokhazadeh, T.; Mollazadeh, R. Sudden cardiac death in COVID-19 patients, a report of three cases. *Future Cardiol.* **2021**, *17*, 113–118. [CrossRef]
75. Gullach, A.J.; Risgaard, B.; Lynge, T.H.; Jabbari, R.; Glinge, C.; Haunsø, S.; Backer, V.; Winkel, B.G.; Tfelt-Hansen, J. Sudden death in young persons with uncontrolled asthma—A nationwide cohort study in Denmark. *BMC Pulm. Med.* **2015**, *15*, 1–8. [CrossRef]
76. Charlier, P.; Naneix, A.; Brun, L.; Álvarez, J.; De La Grandmaison, G.L. Asthma-related sudden death: Clinicopathological features in 14 cases. *Rev. Médecine Légale* **2012**, *3*, 115–119. [CrossRef]
77. Pravettoni, V.; Incorvaia, C. Diagnosis of exercise-induced anaphylaxis: Current insights. *J. Asthma Allergy* **2016**, *9*, 191–198. [CrossRef]
78. Flannagan, L.M.; Wolf, B.C. Sudden death associated with food and exercise. *J. Forensic Sci.* **2004**, *49*, 1–3. [CrossRef]
79. Heim, S.W.; Maughan, K.L. Foreign bodies in the ear, nose, and throat. *Am. Fam. Physician* **2007**, *76*, 121–123.
80. Zhang, L.; Yin, Y.; Zhang, J.; Zhang, H. Removal of foreign bodies in children's airways using flexible bronchoscopic CO2cryotherapy. *Pediatric Pulmonol.* **2016**, *51*, 943–949. [CrossRef] [PubMed]
81. Salih, A.M.; AlFaki, M.; Alam-Elhuda, D.M. Airway foreign bodies: A critical review for a common pediatric emergency. *World J. Emerg. Med.* **2016**, *7*, 5–12. [CrossRef] [PubMed]
82. Japundžić-Žigon, N.; Šarenac, O.; Lozić, M.; Vasić, M.; Tasić, T.; Bajić, D.; Kanjuh, V.; Murphy, D. Sudden death: Neurogenic causes, prediction and prevention. *Eur. J. Prev. Cardiol.* **2018**, *25*, 29–39. [CrossRef]
83. Roberts, C.C.; Snipes, G.J.; Ko, J.M.; Roberts, W.C.; Guileyardo, J.M. Nontraumatic intracerebral hemorrhage unassociated with arterial aneurysmal rupture as a cause of sudden unexpected death. *Bayl. Univ. Med Cent. Proc.* **2014**, *27*, 331–333. [CrossRef] [PubMed]
84. Verdecchia, P.; Angeli, F.; Cavallini, C.; Aita, A.; Turturiello, D.; De Fano, M.; Reboldi, G. Sudden Cardiac Death in Hypertensive Patients. *Hypertension* **2019**, *73*, 1071–1078. [CrossRef]
85. Sessa, F.; Anna, V.; Messina, G.; Cibelli, G.; Monda, V.; Marsala, G.; Ruberto, M.; Biondi, A.; Cascio, O.; Bertozzi, G.; et al. Heart rate variability as predictive factor for sudden cardiac death. *Aging* **2018**, *10*, 166–177. [CrossRef]
86. Algra, A.; Gates, P.C.; Fox, A.J.; Hachinski, V.; Barnett, H.J. Side of Brain Infarction and Long-Term Risk of Sudden Death in Patients with Symptomatic Carotid Disease. *Stroke* **2003**, *34*, 2871–2875. [CrossRef]
87. Lindbohm, J.V.; Kaprio, J.; Jousilahti, P.; Salomaa, V.; Korja, M. Risk Factors of Sudden Death from Subarachnoid Hemorrhage. *Stroke* **2017**, *48*, 2399–2404. [CrossRef] [PubMed]
88. Sharew, A.; Bodilsen, J.; Hansen, B.R.; Nielsen, H.; Brandt, C.T. The cause of death in bacterial meningitis. *BMC Infect. Dis.* **2020**, *20*, 1–9. [CrossRef]
89. Seo, M.H.; Sung, W.Y. A case of near-sudden unexpected death in epilepsy due to ventricular fibrillation. *Open Access Emerg. Med.* **2019**, *11*, 161–166. [CrossRef] [PubMed]
90. Riezzo, I.; Zamparese, R.; Neri, M.; De Stefano, F.; Parente, R.; Pomara, C.; Turillazzi, E.; Ventura, F.; Fineschi, V. Sudden, unexpected death due to glioblastoma: Report of three fatal cases and review of the literature. *Diagn. Pathol.* **2013**, *8*, 73. [CrossRef] [PubMed]
91. Menezes, R.G.; Ahmed, S.; Pasha, S.B.; Hussain, S.A.; Fatima, H.; Kharoshah, M.A.; Madadin, M. Gastrointestinal causes of sudden unexpected death: A review. *Med. Sci. Law* **2017**, *58*, 5–15. [CrossRef]
92. Pomara, C.; Bello, S.; D'Errico, S.; Greco, M.; Fineschi, V. Sudden death due to a dissecting intramural hematoma of the esophagus (DIHE) in a woman with severe neurofibromatosis-related scoliosis. *Forensic Sci. Int.* **2013**, *228*, e71–e75. [CrossRef]

93. Kibayashi, K.; Sumida, T.; Shojo, H.; Tokunaga, O. Unexpected death due to intestinal obstruction by a duplication cyst in an infant. *Forensic Sci. Int.* **2007**, *173*, 175–177. [CrossRef]
94. Lambert, M.P.; Heller, D.; Bethel, C. Extensive Gastric Heterotopia of the Small Intestine Resulting in Massive Gastrointestinal Bleeding, Bowel Perforation, and Death: Report of a Case and Review of the Literature. *Pediatric Dev. Pathol.* **2000**, *3*, 277–280. [CrossRef]
95. Menezes, R.G.; Padubidri, J.R.; Babu, Y.R.; Naik, R.; Kanchan, T.; Senthilkumaran, S.; Chawla, K. Sudden unexpected death due to strangulated inguinal hernia. *Medico-Legal J.* **2016**, *84*, 101–104. [CrossRef]
96. Bernal, W.; Auzinger, G.; Dhawan, A.; Wendon, J. Acute liver failure. *Lancet* **2010**, *376*, 190–201. [CrossRef]
97. Stoppacher, R. Sudden Death Due to Acute Pancreatitis. *Acad. Forensic Pathol.* **2018**, *8*, 239–255. [CrossRef] [PubMed]
98. Palmiere, C. Sudden death due to acute adrenal crisis. *Forensic Sci. Med. Pathol.* **2015**, *11*, 629. [CrossRef] [PubMed]
99. D'Errico, S.; Pomara, C.; Riezzo, I.; Neri, M.; Turillazzi, E.; Fineschi, V. Cardiac failure due to epinephrine-secreting pheochromocytoma: Clinical, laboratory and pathological findings in a sudden death. *Forensic Sci. Int.* **2009**, *187*, e13–e17. [CrossRef]
100. Kjærulff, M.L.B.G.; Astrup, B.S. Sudden death due to diabetic ketoacidosis following power failure of an insulin pump: Autopsy and pump data. *J. Forensic Leg. Med.* **2019**, *63*, 34–39. [CrossRef] [PubMed]
101. Quirante, F.A.P.; Pastor-Pérez, F.J.; Manzano-Fernández, S.; Rivas, N.L.; Pérez, P.P.; Hernández, J.P.; Blanes, J.G. Unexpected autopsy findings after sudden cardiac death: Cardiovascular myxoedema and endocardial fibroelastosis. *Int. J. Cardiol.* **2015**, *182*, 281–283. [CrossRef]
102. Kiewiet, R.M.; Ponssen, H.H.; Janssens, E.N.W.; Fels, P.W. Ventricular fibrillation in hypercalcaemic crisis due to primary hyperparathyroidism. *Neth. J. Med.* **2004**, *62*, 94–96. [PubMed]
103. Wu, L.-S.; Wu, C.-T.; Hsu, L.-A.; Luqman, N.; Kuo, C.-T. Brugada-like electrocardiographic pattern and ventricular fibrillation in a patient with primary hyperparathyroidism. *Europace* **2007**, *9*, 172–174. [CrossRef]
104. Di Nunno, N.; Patanè, F.G.; Amico, F.; Asmundo, A.; Pomara, C. The Role of a Good Quality Autopsy in Pediatric Malpractice Claim: A Case Report of an Unexpected Death in an Undiagnosed Thymoma. *Front. Pediatr.* **2020**, *8*. [CrossRef]
105. Peterslund, P.; Elvander, C.F.; Hoffmann-Peterssen, J. Iatrogenic clinically cardiac arrest after administration of nitroglycerin. *Ugeskr. Laeger* **2019**, *181*, V01190015.
106. Wheeler, S.J.; Wheeler, D.W. Medication errors in anaesthesia and critical care. *Anaesthesia* **2005**, *60*, 257–273. [CrossRef]
107. Sicouri, S.; Antzelevitch, C. Mechanisms Underlying the Actions of Antidepressant and Antipsychotic Drugs That Cause Sudden Cardiac Arrest. *Arrhythmia Electrophysiol. Rev.* **2018**, *7*, 199–209. [CrossRef] [PubMed]
108. Sessa, F.; Salerno, M.; Di Mizio, G.; Bertozzi, G.; Messina, G.; Tomaiuolo, B.; Pisanelli, D.; Maglietta, F.; Ricci, P.; Pomara, C. Anabolic Androgenic Steroids: Searching New Molecular Biomarkers. *Front. Pharmacol.* **2018**, *9*, 9. [CrossRef] [PubMed]
109. Sessa, F.; Salerno, M.; Cipolloni, L.; Bertozzi, G.; Messina, G.; Di Mizio, G.; Asmundo, A.; Pomara, C. Anabolic-androgenic steroids and brain injury: miRNA evaluation in users compared to cocaine abusers and elderly people. *Aging* **2020**, *12*, 15314–15327. [CrossRef]
110. Fineschi, V.; Baroldi, G.; Monciotti, F.; Reattelli, L.P.; Turillazzi, E. Anabolic steroid abuse and cardiac sudden death: A pathologic study. *Arch. Pathol. Lab. Med.* **2001**, *125*, 253–255. [CrossRef]
111. Torrisi, M.; Pennisi, G.; Russo, I.; Amico, F.; Esposito, M.; Liberto, A.; Cocimano, G.; Salerno, M.; Rosi, G.L.; Di Nunno, N.; et al. Sudden Cardiac Death in Anabolic-Androgenic Steroid Users: A Literature Review. *Medicina* **2020**, *56*, 587. [CrossRef] [PubMed]
112. Turillazzi, E.; Bello, S.; Neri, M.; Pomara, C.; Riezzo, I.; Fineschi, V. Cardiovascular effects of cocaine: Cellular, ionic and molecular mechanisms. *Curr. Med. Chem.* **2012**, *19*, 5664–5676. [CrossRef]
113. Lingamfelter, D.C.; Knight, L.D. Sudden death from massive gastrointestinal hemorrhage associated with crack cocaine use: Case report and review of the literature. *Am. J. Forensic Med. Pathol.* **2010**, *31*, 98–99. [CrossRef]
114. Akinlonu, A.; Suri, R.; Yerragorla, P.; López, P.D.; Mene-Afejuku, T.O.; Ola, O.; Dumancas, C.; Chalabi, J.; Pekler, G.; Visco, F.; et al. Brugada Phenocopy Induced by Recreational Drug Use. *Case Rep. Cardiol.* **2018**, *2018*, 1–4. [CrossRef]
115. Stockholm, S.C.; Rosenblum, A.; Byrd, A.; Mery-Fernandez, E.; Bhandari, M. Cannabinoid-Induced Brugada Syndrome: A Case Report. *Cureus* **2020**, *12*, e8615. [CrossRef]
116. Boland, D.M.; Reidy, L.J.; Seither, J.M.; Radtke, J.M.; Lew, E.O. Forty-Three Fatalities Involving the Synthetic Cannabinoid, 5-Fluoro- ADB: Forensic Pathology and Toxicology Implications. *J. Forensic Sci.* **2020**, *65*, 170–182. [CrossRef]
117. Morentin, B.; Callado, L.F. Sudden cardiac death associated to substances of abuse and psychotropic drugs consumed by young people: A population study based on forensic autopsies. *Drug Alcohol Depend.* **2019**, *201*, 23–28. [CrossRef] [PubMed]
118. Neri, M.; Othman, S.M.; Cantatore, S.; De Carlo, D.; Pomara, C.; Riezzo, I.; Turillazzi, E.; Fineschi, V. Sudden infant death in an 8-month-old baby with dengue virus infection: Searching for virus in postmortem tissues by immunohistochemistry and Western blotting. *Pediatric Infect. Dis. J.* **2012**, *31*, 878–880. [CrossRef] [PubMed]
119. Sorkin, T.; Sheppard, M.N. Sudden unexplained death in alcohol misuse (SUDAM) patients have different characteristics to those who died from sudden arrhythmic death syndrome (SADS). *Forensic Sci. Med. Pathol.* **2017**, *13*, 278–283. [CrossRef]
120. Goldstein, R.D.; Blair, P.; Sens, M.A.; Shapiro-Mendoza, C.K.; Krous, H.F.; Rognum, T.O.; Moon, R.Y. Inconsistent classification of unexplained sudden deaths in infants and children hinders surveillance, prevention and research: Recommendations from The 3rd International Congress on Sudden Infant and Child Death. *Forensic Sci. Med. Pathol.* **2019**, *15*, 622–628. [CrossRef]
121. Bolliger, S.A.; Thali, M.J. Imaging and virtual autopsy: Looking back and forward. *Philos. Trans. R. Soc. B Biol. Sci.* **2015**, *370*, 20140253. [CrossRef]

122. Ruder, T.; Thali, M.; Hatch, G. Essentials of forensic post-mortem MR imaging in adults. *Br. J. Radiol.* **2014**, *87*, 20130567. [CrossRef]
123. Bello, S.; Neri, M.; Grilli, G.; Pascale, N.; Pomara, C.; Riezzo, I.; Turillazzi, E.; Fineschi, V. Multi-phase post-mortem CT-angiography (MPMCTA) is a very significant tool to explain cardiovascular pathologies. A sudden cardiac death case. *Exp. Clin. Cardiol.* **2014**, *13*, 855–865.
124. Kornegoor, R.; A A Brosens, L.; Roothaan, S.M.; Smits, A.J.J.; Vink, A. Digitalization of post-mortem coronary angiography. *Histopathol.* **2009**, *55*, 760–761. [CrossRef]
125. Grabherr, S.; Grimm, J.; Heinemann, A. *Atlas of Postmortem Angiography*; Springer: London, UK, 2016.
126. Michaud, K.; Grabherr, S.; Jackowski, C.; Bollmann, M.D.; Doenz, F.; Mangin, P. Postmortem imaging of sudden cardiac death. *Int. J. Leg. Med.* **2014**, *128*, 127–137. [CrossRef] [PubMed]
127. Michaud, K.; Genet, P.; Sabatasso, S.; Grabherr, S. Postmortem imaging as a complementary tool for the investigation of cardiac death. *Forensic Sci. Res.* **2019**, *4*, 211–222. [CrossRef]
128. Hueck, U.; Muggenthaler, H.; Hubig, M.; Heinrich, A.; Güttler, F.; Wagner, R.; Mall, G.; Teichgräber, U. Forensic postmortem computed tomography in suspected unnatural adult deaths. *Eur. J. Radiol.* **2020**, *132*, 109297. [CrossRef]
129. Barcelo, B.; Noce, V.; Gomila, I.; Martín, B.B. Building Bridges between Clinical and Forensic Toxicology Laboratories. *Curr. Pharm. Biotechnol.* **2018**, *19*, 99–112. [CrossRef]
130. Bjune, T.; Risgaard, B.; Kruckow, L.; Glinge, C.; Ingemann-Hansen, O.; Leth, P.M.; Linnet, K.; Banner, J.; Winkel, B.G.; Tfelt-Hansen, J. Post-mortem toxicology in young sudden cardiac death victims: A nationwide cohort study. *Europace* **2017**, *20*, 614–621. [CrossRef]
131. Morentín, B.; Callado, L.F.; García-Hernández, S.; Bodegas, A.; Lucena, J. The role of toxic substances in sudden cardiac death. *Span. J. Leg. Med.* **2018**, *44*, 13–21. [CrossRef]
132. Mladěnka, P.; Applová, L.; Patočka, J.; Costa, V.M.; Remiao, F.; Pourová, J.; Mladěnka, A.; Karlíčková, J.; Jahodář, L.; Vopršalová, M.; et al. Comprehensive review of cardiovascular toxicity of drugs and related agents. *Med. Res. Rev.* **2018**, *38*, 1332–1403. [CrossRef]
133. Montisci, M.; El Mazloum, R.; Cecchetto, G.; Terranova, C.; Ferrara, S.D.; Thiene, G.; Basso, C. Anabolic androgenic steroids abuse and cardiac death in athletes: Morphological and toxicological findings in four fatal cases. *Forensic Sci. Int.* **2012**, *217*, e13–e18. [CrossRef]
134. Sessa, F.; Messina, G.; Russo, R.; Salerno, M.; Castracani, C.C.; Distefano, A.; Volti, G.L.; Calogero, A.E.; Cannarella, R.; Mongioi', L.M.; et al. Consequences on aging process and human wellness of generation of nitrogen and oxygen species during strenuous exercise. *Aging Male* **2020**, *23*, 14–22. [CrossRef]
135. Francavilla, C.V.; Sessa, F.; Salerno, M.; Albano, G.D.; Villano, I.; Messina, G.; Triolo, F.; Todaro, L.; Ruberto, M.; Marsala, G.; et al. Influence of Football on Physiological Cardiac Indexes in Professional and Young Athletes. *Front. Physiol.* **2018**, *9*, 9. [CrossRef]
136. Giudicessi, J.R.; Ackerman, M.J. Determinants of incomplete penetrance and variable expressivity in heritable cardiac arrhythmia syndromes. *Transl. Res.* **2013**, *161*, 1–14. [CrossRef]
137. Miles, C.J.; Behr, E.R. The role of genetic testing in unexplained sudden death. *Transl. Res.* **2016**, *168*, 59–73. [CrossRef]
138. Wilhelm, M.; Bolliger, S.A.; Bartsch, C.; Fokstuen, S.; Gräni, C.; Martos, V.; Medeiros Domingo, A.; Osculati, A.; Rieubland, C.; Sabatasso, S.; et al. Sudden cardiac death in forensic medicine—Swiss recommendations for a multidisciplinary approach. *Swiss Med. Wkly.* **2015**, *145*, 14129. [CrossRef]
139. Tester, D.J.; Medeiros-Domingo, A.; Will, M.L.; Ackerman, M.J. Unexplained Drownings and the Cardiac Channelopathies: A Molecular Autopsy Series. *Mayo Clin. Proc.* **2011**, *86*, 941–947. [CrossRef]
140. Skinner, J.R.; Crawford, J.; Smith, W.; Aitken, A.; Heaven, D.; Evans, C.-A.; Hayes, I.; Neas, K.R.; Stables, S.; Koelmeyer, T.; et al. Prospective, population-based long QT molecular autopsy study of postmortem negative sudden death in 1 to 40 year olds. *Hear. Rhythm.* **2011**, *8*, 412–419. [CrossRef]
141. Turillazzi, E.; La Rocca, G.; Anzalone, R.; Corrao, S.; Neri, M.; Pomara, C.; Riezzo, I.; Karch, S.B.; Fineschi, V. Heterozygous nonsense SCN5A mutation W822X explains a simultaneous sudden infant death syndrome. *Virchows Arch.* **2008**, *453*, 209–216. [CrossRef]
142. Pomara, C.; D'Errico, S.; Riezzo, I.; De Cillis, G.P.; Fineschi, V. Sudden cardiac death in a child affected by Prader-Willi syndrome. *Int. J. Leg. Med.* **2005**, *119*, 153–157. [CrossRef]
143. Bafunno, V.; Bury, L.; Tiscia, G.L.; Fierro, T.; Favuzzi, G.; Caliandro, R.; Sessa, F.; Grandone, E.; Margaglione, M.; Gresele, P. A novel congenital dysprothrombinemia leading to defective prothrombin maturation. *Thromb. Res.* **2014**, *134*, 1135–1141. [CrossRef]
144. Santacroce, R.; Santoro, R.; Sessa, F.; Iannaccaro, P.; Sarno, M.; Longo, V.; Gallone, A.; Vecchione, G.; Muleo, G.; Margaglione, M. Screening of mutations of hemophilia A in 40 Italian patients: A novel G-to-A mutation in intron 10 of the F8 gene as a putative cause of mild hemophilia a in southern Italy. *Blood Coagul. Fibrinolysis* **2008**, *19*, 197–202. [CrossRef]
145. Marey, I.; Fressart, V.; Rambaud, C.; Fornes, P.; Martin, L.; Grotto, S.; Alembik, Y.; Gorka, H.; Millat, G.; Gandjbakhch, E.; et al. Clinical impact of post-mortem genetic testing in cardiac death and cardiomyopathy. *Open Med.* **2020**, *15*, 435–446. [CrossRef]
146. Brion, M.; Quintela, I.; Sobrino, B.; Torres, M.; Allegue, C.; Carracedo, A. New technologies in the genetic approach to sudden cardiac death in the young. *Forensic Sci. Int.* **2010**, *203*, 15–24. [CrossRef]

147. Stattin, E.-L.; Westin, I.M.; Cederquist, K.; Jonasson, J.; Jonsson, B.-A.; Mörner, S.; Norberg, A.; Krantz, P.; Wisten, A. Genetic screening in sudden cardiac death in the young can save future lives. *Int. J. Leg. Med.* **2016**, *130*, 59–66. [CrossRef] [PubMed]
148. Khera, A.V.; Mason-Suares, H.; Brockman, D.; Wang, M.; VanDenburgh, M.J.; Senol-Cosar, O.; Patterson, C.; Newton-Cheh, C.; Zekavat, S.M.; Pester, J.; et al. Rare Genetic Variants Associated with Sudden Cardiac Death in Adults. *J. Am. Coll. Cardiol.* **2019**, *74*, 2623–2634. [CrossRef] [PubMed]
149. Morini, E.; Sangiuolo, F.; Caporossi, D.; Novelli, G.; Amati, F. Application of Next Generation Sequencing for personalized medicine for sudden cardiac death (SCD). *Front. Genet.* **2015**, *6*, 55. [CrossRef]
150. Carturan, E.; Tester, D.J.; Brost, B.C.; Basso, C.; Thiene, G.; Ackerman, M.J. Postmortem genetic testing for conventional autopsy-negative sudden unexplained death: An evaluation of different DNA extraction protocols and the feasibility of mutational analysis from archival paraffin-embedded heart tissue. *Am. J. Clin. Pathol.* **2008**, *129*, 391–397. [CrossRef] [PubMed]
151. Basso, C.; Calabrese, F.; Angelini, A.; Carturan, E.; Thiene, G. Classification and histological, immunohistochemical, and molecular diagnosis of inflammatory myocardial disease. *Heart Fail. Rev.* **2012**, *18*, 673–681. [CrossRef]
152. Boczek, N.J.; Tester, D.J.; Ackerman, M.J. The molecular autopsy: An indispensable step following sudden cardiac death in the young? *Herzschrittmachertherapie Elektrophysiologie* **2012**, *23*, 167–173. [CrossRef]
153. Basso, C.; Carturan, E.; Pilichou, K.; Rizzo, S.; Corrado, D.; Thiene, G. Sudden cardiac death with normal heart: Molecular autopsy. *Cardiovasc. Pathol.* **2010**, *19*, 321–325. [CrossRef] [PubMed]
154. Neri, M.; Frati, A.; Turillazzi, E.; Cantatore, S.; Cipolloni, L.; Di Paolo, M.; Frati, P.; La Russa, R.; Maiese, A.; Scopetti, M.; et al. Immunohistochemical Evaluation of Aquaporin-4 and its Correlation with CD68, IBA-1, HIF-1α, GFAP, and CD15 Expressions in Fatal Traumatic Brain Injury. *Int. J. Mol. Sci.* **2018**, *19*, 3544. [CrossRef]
155. Dunn, K.E.; Caleshu, C.; Cirino, A.L.; Ho, C.Y.; Ashley, E.A. A clinical approach to inherited hypertrophy: The use of family history in diagnosis, risk assessment, and management. *Circ. Cardiovasc. Genet.* **2013**, *6*, 118–131. [CrossRef]
156. D'Ovidio, C.; Carnevale, A.; Grassi, V.M.; Rosato, E.; Del Olmo, B.; Coll, M.; Campuzano, O.; Iglesias, A.; Brugada, R.; Oliva, A. Sudden death due to catecholaminergic polymorphic ventricular tachycardia following negative stress-test outcome: Genetics and clinical implications. *Forensic Sci. Med. Pathol.* **2017**, *13*, 217–225. [CrossRef]
157. Cunningham, K.S. The Promise of Molecular Autopsy in Forensic Pathology Practice. *Acad. Forensic Pathol.* **2017**, *7*, 551–566. [CrossRef] [PubMed]
158. Lin, Y.; Williams, N.; Wang, D.; Coetzee, W.; Zhou, B.; Eng, L.S.; Um, S.Y.; Bao, R.; Devinsky, O.; McDonald, T.V.; et al. Applying High-Resolution Variant Classification to Cardiac Arrhythmogenic Gene Testing in a Demographically Diverse Cohort of Sudden Unexplained Deaths. *Circ. Cardiovasc. Genet.* **2017**, *10*. [CrossRef] [PubMed]
159. Paratz, E.D.; Rowsell, L.; Zentner, D.; Parsons, S.; Morgan, N.; Thompson, T.; James, P.; Pflaumer, A.; Semsarian, C.; Smith, K.; et al. Cardiac arrest and sudden cardiac death registries: A systematic review of global coverage. *Open Heart* **2020**, *7*, e001195. [CrossRef] [PubMed]

MDPI

Case Report

Sudden Death from Primary Cerebral Melanoma: Clinical Signs and Pathological Observations

Alfonso Maiellaro [1], Antonio Perna [2], Pasquale Giugliano [3], Massimiliano Esposito [4,*] and Giuseppe Vacchiano [5,*]

1 Legal Medicine Department, A. Cardarelli Hospital, 80131 Naples, Italy; alfonso.maiellaro@aocardarelli.it
2 Pathology Unit, Mauro Scarlato Hospital, 84018 Scafati, Italy; antoniopernadoc@libero.it
3 AORN Sant'Anna e San Sebastiano di Caserta, 81100 San Sebastiano, Italy; dottgiugliano@tiscali.it
4 Legal Medicine, Department of Medical, Surgical and Advanced Technologies, "G.F. Ingrassia", University of Catania, 95123 Catania, Italy
5 Department of Law, University of Sannio, 82100 Benevento, Italy
* Correspondence: massimiliano.esposito91@gmail.com (M.E.); vacchiano@unisannio.it (G.V.); Tel.: +39-3409348781 (M.E.); +39-3475386107 (G.V.)

Abstract: Primary cerebral tumors rarely provoke sudden death. The incidence is often underestimated with reported frequencies in the range of 0.02 to 2.1% in medicolegal autopsy series. Furthermore, primary cerebral melanoma is an uncommon neoplasm. It represents approximately 1% of all melanoma cases and 0.07% of all brain tumors. This neoplasm is very aggressive, and its annual incidence is about 1 in 10 million people. In the present study, a 20-year-old male was admitted to hospital with vomiting, headache, paresthesia and aggressive behavior. A computed tomography (CT) scan of the head was performed showing a hyperdense nodule in the right parietal lobe with inflammation of the Silvian fissure. A complete autopsy was performed 48 h after death. A blackish material was displayed on the skull base, and posterior fossa. Microscopic examination diagnosed primary brain melanoma. A systematic review of the literature was also performed where no previous analogous cases were found. The forensic pathologist rarely encounters primary cerebral melanoma, and for these reasons, it seemed appropriate to describe this case as presenting aspecific clinical symptoms and leading to sudden unexpected death. Histopathological observations are reported and discussed to explain this surprising sudden death caused by a primary cerebral melanoma.

Keywords: primary cerebral melanoma; sudden death; clinical signs; pathological observations

Citation: Maiellaro, A.; Perna, A.; Giugliano, P.; Esposito, M.; Vacchiano, G. Sudden Death from Primary Cerebral Melanoma: Clinical Signs and Pathological Observations. *Healthcare* **2021**, *9*, 341. https://doi.org/10.3390/healthcare9030341

Academic Editor: Pedram Sendi

Received: 24 February 2021
Accepted: 15 March 2021
Published: 17 March 2021

Publisher's Note: MDPI stays neutral with regard to jurisdictional claims in published maps and institutional affiliations.

1. Introduction

Sudden unexpected death due to a primary nervous neoplasm is not frequent in medico-legal activity [1] because the incidence of undiagnosed primary central nervous system tumors has decreased with the emergence of advanced neuroimaging and other improved diagnostic techniques. Occasionally, the forensic pathologist can encounter a cerebral neoplasm that had been undiagnosed or not suspected prior to death. Eberhart et al. [2] observed 11 cases of primary central nervous system tumors resulting in sudden death over a period of 20 years (1980 to 1999). Glioblastomas [3–6], oligodendragliomas [7], astrocitomas [8], ependymomas [9], extramedullary plasmacytomas [10], colloid cysts of third ventricle [11,12] and meningiomas [13], are reported in the literature to cause sudden death with vague or short-term symptoms and limited healthcare access. Current studies of sudden death from brain tumors concern glioblastoma multiforme or astrocytomas. In the present study, a systematic review of sudden death and brain tumors from 1980 to 2021 was conducted, and it was concluded that although articles had been analyzed over a period of 40 years, there are still few articles today that report this type of death. We found only one case of sudden death associated with melanoma brain metastases [14] and it seemed appropriate to describe the histopathological manifestations of this surprising case of sudden death due to primary cerebral melanoma. Glitza [15], in an autopsy series, identified up 80% of patients with melanoma with metastatic involvement of the central nervous system (CNS); however, while the cerebral metastases of melanoma are often observed, primary cerebral melanoma is very rare. Several histogenetic theories on the origin of meningeal melanocytes have been put forward [16]. It has been postulated that during embryogenesis, leptomeningeal melanocytes coming from the multipotential cells of the neural crest can develop into mesodermal and neural elements. Primary cerebral melanomas develop once melanocytes become neoplastic.

The aim of this study was to present a case report and a systematic review of this rare pathology, in order to improve the overall knowledge of forensic pathologists on sudden death from brain tumors, especially in the case of cerebral melanoma. A complete autopsy, comparing radiological findings and histological presentation, could provide a comprehensive overview of such sudden deaths.

2. Materials and Methods

2.1. Case Description

A 20-year-old male was admitted to hospital with vomiting and headache; he was a cannabis user and cigarette smoker (40 cigarettes a day). He referred paresthesia in the upper limbs, psychomotor agitation and aggressive behavior. The neurological examination pointed out a palpebral ptosis and a right eye ophthalmoplegia including paralysis of the left facial nerve. A CT scan of the brain was performed and showed a hyperdense nodule in the right parietal lobe and along the Sylvian fissure, the leptomeningeal spaces were described as affected by an inflammatory process (Figure 1). Fifteen minutes after the CT scan, he died. The study of the medical record and the sudden onset of symptoms defined death as "sudden". The accuracy of the autopsy examination, the histological investigation (H&E and immunohistochemistry), attributed his death to a rare form of primary cerebral melanoma.

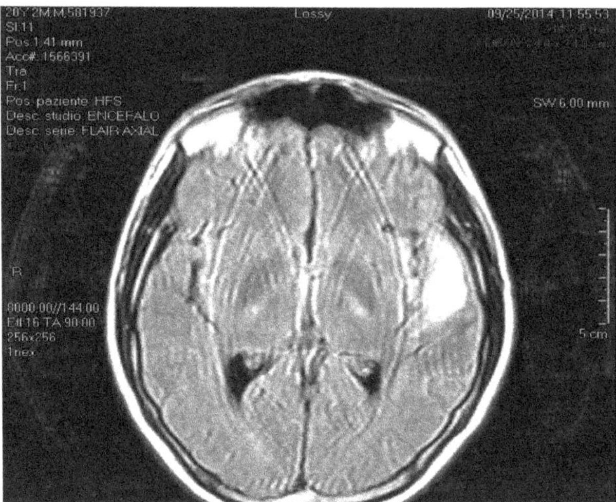

Figure 1. Pre-mortem CT scan of the brain. The right parietal lobe showed inflammatory processes with a hyperdense nodule.

No ethical committee was required. Written informed consent was obtained from his relatives.

2.2. Autopsy Findings

A complete autopsy was performed 48 h after death. External examination revealed abrasions on the hands and the upper limbs. The autopsy revealed the dura was full of clots. The brain weighed 1550 g and was diffusely swollen. In the subarachnoid space of the fronto-parietal lobes and along the interhemispheric cleavage of the parietal left lobe, some nodular blackish rounded soft formations with blood were observed. Moreover, in the temporal and parietal lobes, similar formations were present (Figure 2). A soft and blackish mass (4.5 × 7 cm) enveloped the brain stem and was spreading to the inferior temporal poles and to the cerebellar lobes. There was no herniation of the temporal lobe, unci or cerebellar tonsils. The section of the brain revealed on edematous soft parenchyma. Blackish soft material present in subarachnoid space of the fronto-parietal right lobe invaded the cerebral cortex. The brain ventricles were filled with blood. From the dorsal face of the brain stem, striae of blackish material invaded the nervous tissue (Figure 3). Blackish material was also present along the venous sinus of the skull base and in the posterior fossa (Figure 4). The internal examination of the other organs was unmarkable.

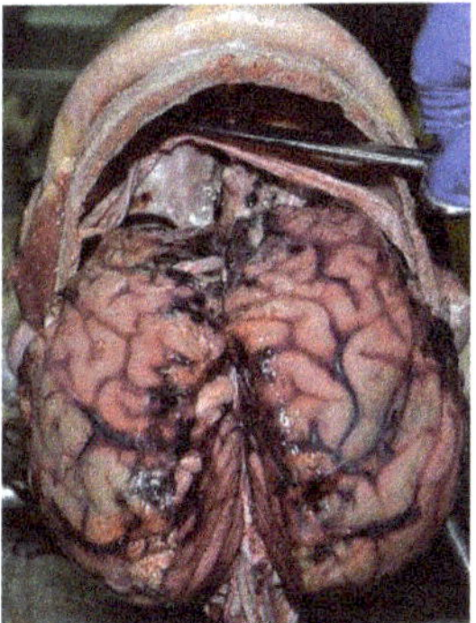

Figure 2. Brain examination. In the temporal and parietal lobes, some nodular blackish rounded soft formations with blood were observed.

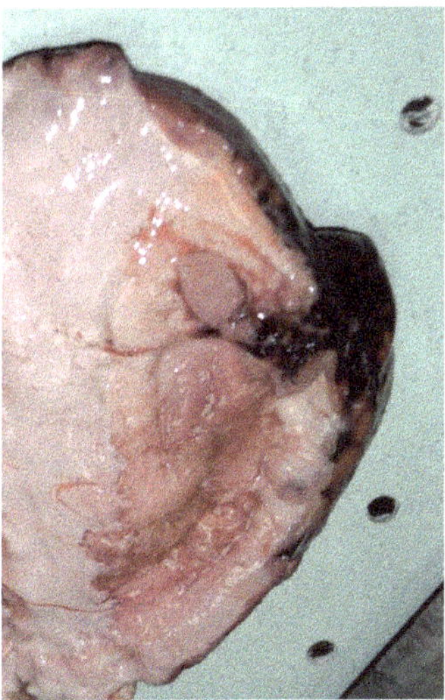

Figure 3. Posterior face of the brain revealed stem striae of blackish material that invaded the nervous tissue.

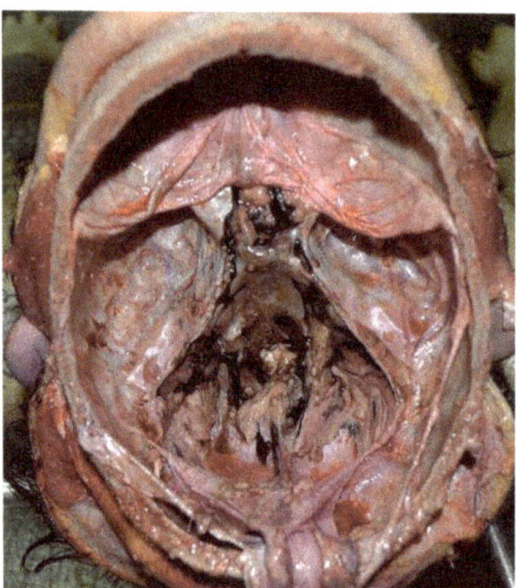

Figure 4. Skull base-posterior fossa and venous sinus examination showed a blackish material.

2.3. *Histological Analysis*

During the autopsy, brain samples were collected that were fixed in 10% buffered formalin. After washing, the water was removed, and they were embedded in paraffin. The obtained blocks were cut (4 µm thickness), using a microtome (Dako, Glostrup, Denmark), sectioned, and stored at room temperature. Sections were stained with Hematoxylin and Eosin (H&E) and Von Gieson's method. The immunohistochemical analysis of the samples was performed through antimelanoma-antibody (HMB45+), Melan A-antibody, CD3-antibody, CD20+-antibody, PenCK-antibody, Ki67+, as suggested by Tosaka et al. [17] for the diagnosis of primary leptomeningeal melanoma. Sections were observed using a Zeiss Axioplan light microscope (Carl Zeiss, Oberkochen, Germany). Subsequently, images were obtained using a Zeiss AxioCam MRc5 digital camera (Carl Zeiss, Oberkochen, Germany). In the present study, the microscopic examination of the brain showed, on the right parietal lobe and in brain-stem tissues, many atypical and pleomorphic cells with voluminous discolored nuclei and melanin-pigment in the cytoplasm. These cells were arranged like a "cordon" or irregularly like a "vortex". Moreover, a vascular proliferation and spotty hemorrhages were also observed. In the cerebral cortex tissues, edema and ischemic dark neurons, and near them, a lymphocytic and macrophage reaction were seen. The leptomeningeal arteriolar vessels showed clots, partially reorganized. Around them, we observed many pleomorphic and atypical cells with melanin pigment in the cytoplasm (Figure 5).

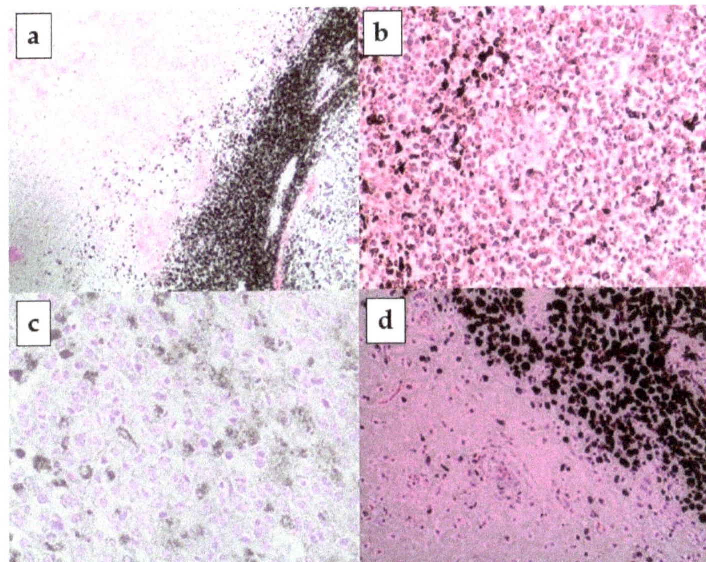

Figure 5. Histological findings. (**a**) neoplastic voluminous cells with melanin in the cerebral tissue (E.E. 60×); (**b**) neoplastic cell population: numerous, voluminous melanoblasts with marked nuclear pleiomorphism; (**c**) neoplastic cells with melanic pigment antimelanoma-antibody (HMB45 + 200×); (**d**) neoplastic cell population near a small cerebral artery (E.E. 200×).

3. Systematic Review

A systematic review was conducted according to the PRISMA guidelines [18].

Pubmed and Google Scholar were used as search engines from 1 January 1980 to 1 February 2021 to evaluate the association between sudden death and brain tumor. meSH was used for the following words: (sudden death) AND (primary cerebral tumor).

3.1. Inclusion and Exclusion Criteria

The following exclusion criteria were used: (1) review, (2) articles not in English, (3) animal studies, (4) abstract, (5) editorial, (6) poster, and (7) communications at conferences. The inclusion criteria were as follows: (1) Original Article, (2) Case Report, and (3) Articles in English.

3.2. Quality Assessment and Data Extraction

A.M. and A.P. initially evaluated all the articles, evaluating the title, the abstract, and the whole text. Once, P.G. and M.E. reanalyzed the articles chosen independently. In cases of conflicting opinions between the articles, they were submitted to G.V.

3.3. Characteristics of Eligible Studies

A total of 388 articles were collected (66 from PubMed and 322 from Google Scholar). Of them, 27 duplicates were removed. A total of 339 articles did not meet the inclusion criteria. In conclusion, 22 articles were included in the present systematic review (Figure 6).

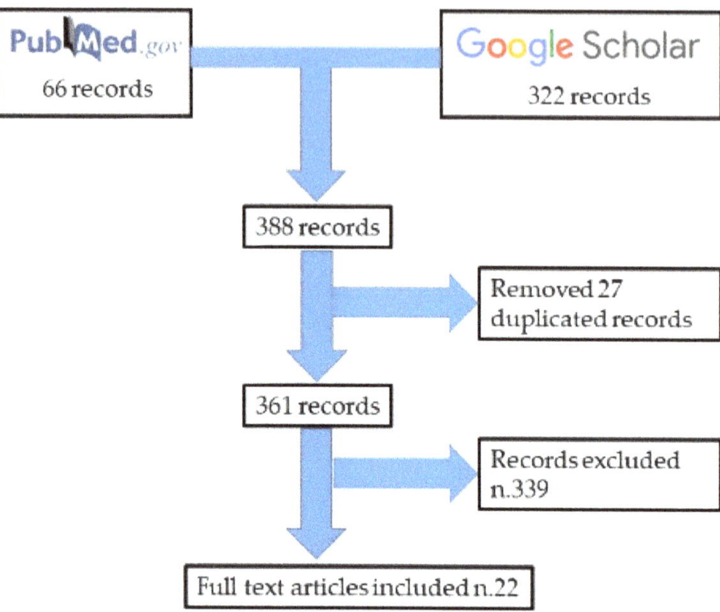

Figure 6. Flow diagram illustrating included and excluded studies in this systematic review.

4. Results

Most of the studies were case reports (n. 16), some cases series (n. 4), and a few original articles (n. 2). Most of the tumors related to sudden death were matched with glioblastoma multiform or astrocytomas in the different world health organization (WHO) grades. Other tumors related to sudden death were oligodendrogliomas, gliomas, adenomas, and colloid cysts. The most frequent autopsy findings were cerebral edema, solid tumor mass, and hemorrhagic infarction of surrounding tissue. The most frequent cause of death was due to intracranial hypertension from the tumor; sometimes it was due to hemorrhage caused by the tumor (especially in cases of glioblastoma multiform); only one case was due to respiratory arrest due to the location of the tumor near the center of breathing control (Table 1).

Table 1. summarizes the details of the systematic review.

Reference	Study Design	Primary Cerebral Tumor	Autopsy Findings	Cause of Death
Eberhart, C.G. [2]	Original Article	Astrocytoma (n.2) Schwannoma (n.1) Glioblastoma multiforme (n.4) Colloid cyst (n.2) Glioma (n.1) Pituitary adenoma (n.1)	Brain edema was shown in all cases. The microscopic study diagnosed the type-specific tumor.	Death was attributed in all cases to hydrocephalus and intracranial hypertension except for one case of glioblastoma multiforme in which death was attributed to massive tumor hemorrhage.
Matschke, J. [3]	Case Series	Glioblastoma multiform (n.3)	Gross examination of the brain showed numerous cystic spaces. Microscopic examination revealed polymorphic astrocytic cells.	Death was attributed to intracranial hypertension.
Gleckman, A.M. [4]	Case Series	Ganglioma (n.1) Astrocytoma (n.1)	Brain edema was shown in all cases. The microscopic study diagnosed the type-specific tumor.	Death was attributed to hydrocephalus and intracranial hypertension.
Sutton, J.T. [5]	Case Report	Glioblastoma multiform	Gross examination of the brain showed a hemorrhagic infiltration of the right lobe equal to $7 \times 5 \times 5$ cm. Microscopic examination revealed hemorrhagic infiltration of the cortex with tumor invasion.	Death was due to an acute hemorrhage of the tumor.
Riezzo, I. [6]	Case Series	Glioblastoma multiforme (n.3)	Macroscopic findings of the brain were characterized by diffuse hypoxia/ischemia and edema of the brain tissue with extensive hemorrhagic infiltration and necrosis confirmed also on histological examination.	Death was attributed in all cases to hydrocephalus and intracranial hypertension.
Manousaki, M. [7]	Case Report	Oligodendroglioma	Brain edema with "fried-egg" cell tumor.	Death was due to hemorrhagic leakage into the cerebrospinal fluid
Vougiouklakis, T. [8]	Case Series	Glioblastoma multiforme (n.1) Astrocytoma WHO grade III (n.1)	The examination of the brain revealed flattening of the fissures with large hemorrhagic infarction in both cases.	Death was due to massive tumor hemorrhage.
Harrison, W.T. [9]	Case Report	Anaplastic Ependymoma	After formalin fixation, the brain showed a $7 \times 6 \times 6$ cm necrotic cavity mass of the lateral ventricle. Microscopically, the tumor has been attributed to an anaplastic ependymoma with parenchyma characterized by fibrillary processes.	Death was attributed to hydrocephalus and intracranial hypertension.
Sidlo, J. [10]	Case Report	Sellar extramedullary plasmacytoma	After brain removal, the examination of the sella turcica showed an intrasellar tumor mass of $5.5 \times 5.5 \times 3$ cm. Histopathological examination showed mature plasma cells with eccentrically positioned round nuclei.	Death was attributed to hydrocephalus and intracranial hypertension

Table 1. *Cont.*

Reference	Study Design	Primary Cerebral Tumor	Autopsy Findings	Cause of Death
Aissaoui, A. [19]	Case Report	Leptomeningeal Melanocytosis	A dark brown mass was present on the basal leptomeninges in the frontal areas. The brain was edematous. Microscopic analysis revealed the dark color of the tumor due to melanin pigments.	Death was attributed to hydrocephalus and intracranial hypertension
Ozkul, A. [20]	Case Report	Leptomeningeal oligodendrogliomatosis	Macroscopic examination revealed edema of the brain. H&E examination showed an invasion of tumor at the brain, cerebellum and spinal cord by plasmacytoid cells.	Death was attributed to hydrocephalus and intracranial hypertension.
Ross, J. [21]	Case Report	Glioma	The brain was characterized by diffuse hypoxia/ischemia and edema of the brain tissue. At H&E examination, a hyper cellularity of glial tumor cells was displayed.	Death was attributed to hydrocephalus and intracranial hypertension
Havlik, D.M. [22]	Case Report	Glioma	Macroscopic examination of the brain showed swelling of hemispheres; at H&E examination, pseudo rosettes and tumor cells were seen.	Death was attributed to intracranial hypertension.
DiMaio, S.M. [23]	Original Article	Colloid cyst (n.1) Oligodendroglioma (n.2) Glioblastoma multiforme (n.2) Astrocytoma WHO grade III (n.3) Medulloblastoma (n.1) Astrocytoma WHO grade II (n.4) Sarcoma (n.1) Teratoma cyst (n.1) Meningioma (n.1) Chromophobe adenoma (n.1)	Brain edema was shown in all cases. The microscopic study diagnosed the type-specific tumor.	Death was attributed in all cases to hydrocephalus and intracranial hypertension.
Lau, G. [24]	Case Report	Intracranial Germinoma	Macroscopic examination was unremarkable with a normal brain size and weight. At microscopic examination the pituitary gland displayed a massive tumor invasion with extensive peripheral lymphoid aggregates.	Death was due to a combination of acute hemorrhage of the tumor combined with microvascular disease of the heart.
Shiferaw, K. [25]	Case Report	Glioblastoma multiform	Macroscopic findings of the brain showed a tumor occupying both frontal lobes with extensive hemorrhagic infiltration and necrosis confirmed also on histological examination.	Death was attributed in all cases to hydrocephalus and intracranial hypertension.
Matturri, L. [26]	Case Report	Hemangioendothelioma	Macroscopic and microscopic examination of the brain found a solid tumor inside the medulla oblongata.	Death was due to the impaired breathing control due to the location of the tumor.

Table 1. *Cont.*

Reference	Study Design	Primary Cerebral Tumor	Autopsy Findings	Cause of Death
Matsumoto, H. [27]	Case Report	Glioblastoma multiform	Macroscopic findings of the brain were characterized by diffuse hypoxia/ischemia and edema of the brain tissue with extensive hemorrhagic infiltration and necrosis confirmed also on histological examination.	Death was attributed in all cases to hydrocephalus and intracranial hypertension.
Prahlow, J.A. [28]	Case Report	Astrocytoma	Macroscopic examination of the brain showed a solid tumor of 1.2 cm. Microscopically, cells of various shapes with nuclei of different size, and microcalcifications of the parenchyma were shown.	Death was due to a seizure disorder related to the tumor combined with acute ethanol intoxication.
Shields, L.B. [29]	Case Report	Pituitary adenoma	The pituitary fossa showed the presence of a red-colored tumor with hemorrhagic infarction. The examination in H&E revealed the presence of tumor with hemorrhagic infiltration.	Death was attributed to intracranial hypertension.
Ortiz-Reyes, R. [30]	Case Report	Subependymoma	Gross examination of the brain showed bilateral ventricular dilatation and inside a tumor of 3 cm in diameter. Microscopic examination showed meningothelial tumoral cells.	Death was due to the impaired breathing control due to the location of the tumor.
Nelson, J. [31]	Case Report	Ganglioma of the medulla	Examination of the brain after fixation showed a medulla with a mass invading the cerebellum. Microscopic examination revealed two types of neoplastic cells: astrocytes and oligodendrocytes.	Death was attributed to intracranial hypertension.

5. Discussion

Sudden death due to brain tumors has a low incidence and is often underestimated [18]. This systematic review highlights that currently there is no study with a large number of cases of sudden death from brain tumors. Furthermore, most of the case reports report glioblastoma multiforme. This case report is the first to focus on sudden death from cerebral melanoma and it is important that forensic pathologists recognize this cause of sudden death.

Primary cerebral melanoma is a very uncommon neoplasm derived from melanocytes present in the leptomeninges. This pathology, first described by Virchow [32], is very rare, with an annual incidence of approximately 1 in 10 million people [33] and represents about 1% of all melanoma cases and 0.07% of all brain tumors [34]. Indeed, many studies have demonstrated that in central nervous systems (CNS) metastases of melanoma are also frequently present because the melanoma has a high risk of spreading to CNSs. The melanoma cells, in fact, share with vascular cells numerous cell surface molecules; they are highly angiogenic and possess a higher degree of "stemness" than other solid tumors [15]. According to Hayword [35], we can identify a primary cerebral melanoma when the following occur: (a) no malignant melanoma outside the CNS; (b) non-malignant neoplasm in other parts of the CNS; (c) histological validation of melanoma.

Age is also an additional factor to distinguish a primary cerebral melanoma from metastatic melanoma. Primary cerebral melanoma develops primarily in patients under 50 years of age and rarely metastasize to other organs. Metastatic melanomas develop mainly in the elderly, have a rapid clinical course and multiple intracerebral diffusion [36]. In the presented case, no other melanomas were detected at the clinical and forensic examinations of the body; a solitary blackish mass (4 × 7 cm) was observed around the brain stem. Leptomeningeal blackish soft rounded formations in the fronto-parietal lobes were observed and the patient was under 50 years old. The histological observations confirmed, with the immunohistochemical analysis, the diagnosis of melanoma. In addition to the blackish mass observed around the brain stem, we also observed in the subarachnoid space in the right parietal lobe some rounded blackish nodules invading the nervous tissue. We interpreted these modifications as the primitive cerebral melanoma located in the brain stem subarachnoid space, where we found a blackish mass at autopsy. The meningeal vessels of the left parietal lobe were thrombosed and around these vessels we observed many atypical pleomorphic cells with cytoplasmatic melanin pigment; the neoplasm had probably first developed from the leptomeningeal space around the brain stem, and after, spread to the subarachnoid space of the left parietal lobe. In fact, it should be borne in mind also that melanocytes far outnumber leptomeninges of the skull base and cervical cord [37,38] and the most common locations of primary cerebral melanoma are the anterolateral face of the spinal cord and the postero-lateral face of the brain stem [39]. In agreement with this author, primary cerebral melanoma was observed also in the posterior fossa in the cerebellopontine angle [40,41] and Arantes [42] reported 13 cases of primary malignant melanoma derived from the pineal body. Greco Castro [43] observed this neoplasm in the temporal lobe near the Sylvian fissure and Quillo-Olvera [44] reported that the primary cerebral melanoma can be found frequently in the cerebral lobe (53.1%) in the posterior fossa (17.3%) and pineal region (13.6%). We believe that primary cerebral melanoma may occur in any leptomeningeal location, although our reported case is in agreement with the observations of Troya-Castilla. The clinical manifestations are generally non-specific, related to an increased cerebral pressure, or cerebral hemorrhages, or a neurological dysfunction and the early diagnosis of this neoplasm is a very difficult challenge for clinicians [45–48]. The aggressive behavior, the vague symptoms, and the unspecific neuroimaging make the prognosis of this tumor very difficult [49]. Several studies [50,51] reported a survival of ~4 months from diagnosis of cerebral nervous system melanoma, and once the cancer spreads to the leptomeninges, the overall median survival is generally only 10 weeks [52,53]. In our case, two months after the first symptoms, the patient died. The patient had a history of drug addiction, presenting psychomotor agitation and aggressive behavior, as can be observed in many cerebral neoplasms [54–57]. Morais described a case of delirium in a patient with brain melanoma metastases [58] and other authors reported headache and diplopia in a 27-year-old man affected by primary meningeal melanocytoma in the anterior cranial fossa [59]. However, we can ascribe the first presented symptoms (headache and vomit) to increased intracranial pressure, probably due to the increased neoplastic mass in the posterior fossa and a subarachnoid hemorrhage. However, the symptoms observed before death (paralysis of the III and VIII cranial nerve) clearly showed a suffering brain stem. The involvement of this important brain structure also explains the sudden death due to an alteration of the hypothalamic cardiovascular regulatory centers. It is well-known that stimulation of the hypothalamus can lead to anatomic cardiovascular disturbances and cardiac arrest [60,61].

6. Conclusions

The surprising case of sudden death stresses the need for a careful analysis in all patients with psychiatric disorders or no-specific neurological signs, especially when these manifestations show-up unexpectedly in a healthy person. Despite the fact that modern genetic [62–64], radiological [65–67], and therapeutic treatments [68–73] have revolutionized the approach to primary cerebral melanoma, autopsy can still provide a useful support in determining the exact location, diffusion, and histological pattern of this neoplasm, and the cause of death. In human beings, melanocyte exists in the uvea, cerebral parenchyma, leptomeninges, and mucous membranes. Primary cerebral melanoma is rare and has an estimated incidence of 0.005 cases per 100,000 and develops in around 1% of all melanoma cases. Brain melanoma metastasis is much more common than primary brain melanoma. Primary brain melanomas originate from melanocytes of the leptomeninges and histopathologically show strongly pigmented cells with prominent nucleoli [74]. Through this systematic review, the present case is the first described in literature in which a primary brain melanoma caused a sudden death. It is, therefore, of crucial importance to report this case to the scientific community.

Author Contributions: Conceptualization, A.M. and A.P.; methodology, P.G.; software, M.E.; validation, G.V., A.M. and A.P.; formal analysis, P.G.; investigation, M.E.; resources, G.V.; data curation, A.M.; writing—original draft preparation, A.P.; writing—review and editing, M.E.; visualization, G.V.; supervision, P.G.; project administration, A.M.; funding acquisition, G.V. All authors have read and agreed to the published version of the manuscript.

Funding: This research received no external funding.

Institutional Review Board Statement: All procedures performed in the study were in accordance with the ethical standards of the institution and with the 1964 Helsinki Declaration and its later amendments or comparable ethical standards.

Informed Consent Statement: Informed consent was obtained from the relatives.

Data Availability Statement: All data are included in the main text.

Acknowledgments: The authors thank the Scientific Bureau of the University of Catania for language support.

Conflicts of Interest: The authors declare no conflict of interest.

Ethical Approval and Consent to Participate: All procedures performed in the study were in accordance with the ethical standards of the institution and with the 1964 Helsinki Declaration and its later amendments or comparable ethical standards. Informed consent was obtained from relatives.

References

1. Back, M.; Graham, D.I. Sudden unexplained death in adults caused by intracranial pathology. *J. Clin. Pathol.* **2002**, *55*, 44–50.
2. Eberhart, C.G.; Morrison, A.; Gyure, K.A.; Frazier, J.; Smialek, J.E.; Troncoso, J.C. Drecreasing incidence of sudden death due to undiagnosed primary central nervous system tumors. *Arch. Pathol. Lab. Med.* **2001**, *125*, 1024–1030. [CrossRef] [PubMed]
3. Matschke, J.; Tsokos, M. Sudden unexpected death due to undiagnosed glioblastoma: Report of three cases and review of the literature. *Int. J. Leg. Med.* **2005**, *119*, 280–284. [CrossRef]

4. Gleckman, A.M.; Smith, T.W. Unexpected death from primary posterior fossa tumors. *Am. J. Forensic Med. Pathol.* **1998**, *19*, 303–308. [CrossRef] [PubMed]
5. Sutton, J.T.; Cummings, P.M.; Ross, G.W.; Lopes, M.B. Sudden death of a 7-year-old boy due to undiagnosed glioblastoma. *Am. J. Forensic Med. Pathol.* **2010**, *31*, 278–280. [CrossRef] [PubMed]
6. Riezzo, I.; Zamparese, R.; Neri, M.; De Stefano, F.; Parente, R.; Pomara, C.; Turillazzi, E.; Ventura, F.; Fineschi, V. Sudden, unexpected death due to glioblastoma: Report of three fatal cases and review of literature. *Diagn. Pathol.* **2013**, *8*, 73. [CrossRef]
7. Manousaki, M.; Papadaki, H.; Papavdi, A.; Kranioti, E.F.; Mylonakis, P.; Varakis, S.; Michalodimitrakis, M. Sudden unexpected death from oligodendroglioma: A case report and review of the literature. *Am. J. Forensic Med. Pathol.* **2011**, *32*, 336–340. [CrossRef] [PubMed]
8. Vougiouklakis, T.; Mitselou, A.; Agnantis, N.J. Sudden death due to primary intracranial neoplasm—A forensic autopsy study. *Anticancer Res.* **2006**, *26*, 2463–2466. [PubMed]
9. Harrison, W.T.; Bouldin, T.; Buckley, A.; Estrada, J.; McLondon, R.; Jansen, K. Sudden unexpected death in a child from anaplastic ependymoma. *Am. J. Forensic Med. Pathol.* **2019**, *40*, 275–278. [CrossRef] [PubMed]
10. Sidlo, J.; Sidlova, H. Sudden and unexpected death due to intracranial sellar extramedullary plasmacytoma. *J. Forensic Leg. Med.* **2019**, *61*, 89–91. [CrossRef] [PubMed]
11. Cuoco, J.A.; Rogers, C.M.; Busch, C.M.; Benko, M.J.; Apfel, L.S.; Elias, Z. Postexercise death due to hemorrhagic colloid cyst of third ventricle: Case report and literature review. *World Neurosurg.* **2019**, *123*, 351–356. [CrossRef] [PubMed]
12. Turillazzi, E.; Bello, S.; Neri, M.; Riezzo, I.; Fineschi, V. Colloid cyst of the third ventricle, hypothalamus, and heart: A dangerous link for sudden death. *Diagn. Pathol.* **2012**, *7*, 144–148. [CrossRef]
13. Gitto, L.; Bolino, G.; Cina, S.J. Sudden unexpected deaths due to intracranial meningioma: Presentation of six fatal cases, review of the literature and a discussion of the mechanism of death. *J. Forensic Sci.* **2018**, *63*, 947–953. [CrossRef] [PubMed]
14. Durao, H.C.; Veiga, G.M.; Pedro, N.; Goncalves, M.M.; Pedrosa, F. Sudden death associated with melanoma brain metastates. *J. Pathol. Nepal* **2018**, *8*, 1412–1415.
15. Glitza, I.C.; Heimberger, A.B.; Sulman, E.P.; Davies, M.A. Prognostic factors for survival in melanoma patients with brain metastates. In *Brain Metastates from Primary Tumors*; Hayat, M., Ed.; Academic Press: New York, NY, USA, 2016; pp. 267–292.
16. Suranagi, V.V.; Maste, P.; Malur, P.R. Primary intracranial malignant melanoma: A rare case with review of literature. *Asian J. Neurosurg.* **2015**, *10*, 39–41. [CrossRef] [PubMed]
17. Tosaka, M.; Tamura, M.; Oriuchi, N. Cerebrospinal fluid immunocytochemical analysis and neuroimaging in the diagnosis of primary leptomenigeal melanoma: Case report. *J. Neurosurg.* **2001**, *94*, 528–532. [CrossRef]
18. Moher, D.; Shamseer, L.; Clarke, M.; Ghersi, D.; Liberati, A.; Petticrew, M.; Shekelle, P.; Stewart, L.A. Preferred reporting items for systematic review and meta-analysis protocols (PRISMA-P) 2015 statement. *Syst. Rev.* **2015**, *4*, 1. [CrossRef] [PubMed]
19. Aissaoui, A.; Mosrati, M.A.; Moussa, A.; Belhaj, M.; Bougattas, M.; Zakhama, A.; Chadly, A. Sudden death and primary leptomeningeal melanocytosis: A case report with an autopsy diagnosis. *Am. J. Forensic Med. Pathol.* **2015**, *36*, 199–201. [CrossRef] [PubMed]
20. Ozkul, A.; Meteoglu, I.; Tataroglu, C.; Akyol, A. Primary diffuse leptomeningeal oligodendrogliomatosis causing sudden death. *J. Neurooncol.* **2007**, *81*, 75–79. [CrossRef] [PubMed]
21. Ross, J.; Olar, A.; Fuller, C. A Pediatric Case of Diffuse Glioma Diagnosed at Autopsy. *Acad. Forensic Pathol.* **2017**, *7*, 657–666. [CrossRef] [PubMed]
22. Havlik, D.M.; Becher, M.W.; Nolte, K.B. Sudden death due to primary diffuse leptomeningeal gliomatosis. *J. Forensic Sci.* **2001**, *46*, 392–395. [CrossRef] [PubMed]
23. DiMaio, S.M.; DiMaio, V.J.; Kirkpatrick, J.B. Sudden, unexpected deaths due to primary intracranial neoplasms. *Am. J. Forensic Med. Pathol.* **1980**, *1*, 29–45. [CrossRef] [PubMed]
24. Lau, G.; Sng, I. A case of sudden death from primary intracranial germinoma complicated by microvascular disease of the heart. *Forensic Sci. Int.* **2003**, *137*, 1–5. [CrossRef]
25. Shiferaw, K.; Pizzolato, G.P.; Perret, G.; La Harpe, R. Sudden, unexpected death due to undiagnosed frontal glioblastoma in a schizophrenic patient. *Forensic Sci. Int.* **2006**, *158*, 200–203. [CrossRef] [PubMed]
26. Matturri, L.; Ottaviani, G.; Rossi, L. Sudden and unexpected infant death due to an hemangioendothelioma located in the medulla oblongata. *Adv. Clin. Pathol.* **1999**, *3*, 29–33.
27. Matsumoto, H.; Yamamoto, K. A case of sudden death by undiagnosed glioblastoma multiforme. *Nihon Hoigaku Zasshi Jpn. J. Leg. Med.* **1993**, *47*, 336–339.
28. Prahlow, J.A.; Teot, L.A.; Lantz, P.E.; Stanton, C.A. Sudden death in epilepsy due to an isolated subependymal giant cell astrocytoma of the septum pellucidum. *Am. J. Forensic Med. Pathol.* **1995**, *16*, 30–37. [CrossRef]
29. Shields, L.B.; Balko, M.G.; Hunsaker, J.C. Sudden and unexpected death from pituitary tumor apoplexy. *J. Forensic Sci.* **2012**, *57*, 262–266. [CrossRef] [PubMed]
30. Ortiz-Reyes, R.; Dragovic, L.; Eriksson, A. Sudden unexpected death resulting from previously nonsymptomatic subependymoma. *Am. J. Forensic Med. Pathol.* **2002**, *23*, 63–67. [CrossRef] [PubMed]
31. Nelson, J.; Frost, J.L.; Schochet, S.S., Jr. Sudden, unexpected death in a 5-year-old boy with an unusual primary intracranial neoplasm. Ganglioglioma of the medulla. *Am. J. Forensic Med. Pathol.* **1987**, *8*, 148–152. [CrossRef]
32. Virchow, R. Pigment und diffuse melanoses der arachnoides. *Virchows Arch.* **1859**, *16*, 180–182. [CrossRef]

33. Tang, K.; Kong, X.; Mao, G.; Quin, M.; Znov, H.; Zhou, L.; Nie, Q.; Xu, Y.; Du, S. Primary cerebral malignant melanoma. A case report with literature review. *Medicine* **2017**, *96*, e5805. [CrossRef]
34. Liubinas, S.V.; Maartens, N.; Drummond, K.J. Primary melanocytic neoplasm of the central nervous system. *J. Clin. Neurosci.* **2010**, *17*, 1227–1232. [CrossRef]
35. Hayward, R.D. Malignant melanoma and cerebral nervous system. A guide for classification based on the clinical findings. *J. Neurol. Neurosurg. Psychiatry* **1976**, *39*, 526–530. [CrossRef] [PubMed]
36. Gempt, J.; Buchmann, N.; Grams, A.E. Black brain: Transformation of a melanocytoma with diffuse melanocytosis into a primary cerebral melanoma. *J. Neurooncol.* **2011**, *102*, 323–328. [CrossRef] [PubMed]
37. Bydon, A.; Guttierez, J.A.; Mahmood, A. Meningeal melanocytoma: An aggressive course for a benign tumor. *J. Neurooncol.* **2003**, *64*, 259–263. [CrossRef] [PubMed]
38. Rades, D.; Schild, S.E. Dose responde relationship for fractionated irradiation in the treatment of spinal meningeal melanocytomas: A review of the literature. *J. Neurooncol.* **2006**, *77*, 311–314. [CrossRef] [PubMed]
39. Troya-Castilla, M.; Rocha-Romero, S.; Crochon-Gonzales, Y.; Marquez-Rivas, F.J. Primary cerebral malignant melanoma in insular region with extracranial metastasis: Case report and review literature. *World J. Surg. Oncol.* **2016**, *14*, 235–242. [CrossRef] [PubMed]
40. Brachmann, D.E.; Doherty, J.K. CPA melanoma: Diagnosis and management. *Otol. Neurotol.* **2007**, *28*, 529–537. [CrossRef] [PubMed]
41. Bhandari, L.; Alapatt, J.; Govindan-Sreekumar, T. Primary cerebellopontine angle melanoma: A case report and review. *Turk. Neurosurg.* **2012**, *22*, 469–474. [CrossRef]
42. Arantes, M.; Castro, A.F.; Romao, H.; Meireles, P.; Garcia, R.; Honavar, M.; Vaz Ar Resende, M. Primary pineal malignant melanoma: Case report and literature review. *Clin. Neurosurg.* **2011**, *113*, 59–64. [CrossRef] [PubMed]
43. Greco Castro, S.; Soffietti, R.; Bradac, G.B.; Boldorini, R. Primitive cerebral melanoma: Case report and review of the literature. *Surg. Neurol.* **2001**, *55*, 163–168. [CrossRef]
44. Quillo-Olvera, J.; Uribe-Olalde, J.S.; Alcantara-Gomez, L.A.; Rejon-Perez, J.D.; Palomera-Gomez, H.G. Primary malignant melanoma of the central nervous system: A diagnostic challenge. *Cir. Cir.* **2015**, *83*, 129–134. [CrossRef] [PubMed]
45. Wadasadawala, T.; Trivedi, S.; Gupta, T.; Epari, S.; Jalari, R. The diagnostic dilemma of primary central nervous system melanoma. *J. Clin. Neurosci.* **2010**, *17*, 1014–1017. [CrossRef] [PubMed]
46. Lee, P.H.; Wang, L.C.; Lee, E.J. Primary intracranial melanoma. *J. Cancer Res. Pract.* **2017**, *4*, 23–26. [CrossRef]
47. Berzero, G.; Diamanti, L.; Di Stefano, A.L.; Bini, P.; Franciotta, D.; Imarisio, I.; Pedrazzoli, P.; Magrassi, L.; Morbini, P.; Farina, L.M.; et al. Meningeal Melanomatosis: A challenge for timely diagnosis. *BioMed Res. Int.* **2015**, *2015*, 6. [CrossRef]
48. Gleissner, B.; Chamberlain, M.C. Neoplastic meningitis. *Lancet Neurol.* **2006**, *5*, 443–452. [CrossRef]
49. Sarnast, A.H.; Mujtaba, B.; Bhat, A.R.; Kirmani, A.R.; Tanki, H.N. A unique case of primary intracranial melanoma. *Asian J. Neurosurg.* **2018**, *13*, 168–171. [CrossRef] [PubMed]
50. Davies, M.A.; Liu, P.; Mcintyre, S.; Kim, K.B.; Papadopoulos, N.; Hwu, W.J.; Bedikian, A. Prognostic factors for survival in melanoma patients with brain metastates. *Cancer* **2011**, *117*, 1687–1696. [CrossRef] [PubMed]
51. Raizer, J.J.; Hwu, W.J.; Panageas, K.S. Brain and leptomenigeal metastates from cutaneous melanoma: Survival outcomes based on clinical features. *Neurol. Oncol.* **2008**, *10*, 199–207. [CrossRef]
52. Cohen, J.V.; Tawbi, H.; Margolin, K.A.; Amravadi, R.; Bosenberg, M.; Brastianos, P.K.; Chiang, V.L.; De Groot, J.; Glitza, I.C.; Herlin, M. Melanoma central nervous system metastates: Current approaches, challenges, and opportunities. *Pigment Cell Melanoma Res.* **2016**, *29*, 627–642. [CrossRef] [PubMed]
53. Cohen, J.V.; Alomari, A.K.; Vortmeyer, A.O.; Jilaveanu, L.B.; Gold-Berg, S.B.; Mahajan, A.; Chiang, V.L.; Kluger, H.M. Melanoma brain metastasis pseudoprogression after pembrolizumab treatment. *Cancer Immunol. Res.* **2016**, *4*, 179–182. [CrossRef] [PubMed]
54. Braeuer, R.R.; Watson, I.R.; Wu, C.S.; Mobley, A.K.; Kamiya, T.; Shoshan, E.; Bar-Eli, M. Why is melanoma so metastatic? *Pigment Cell Melanoma Res.* **2013**, *27*, 19–36. [CrossRef]
55. Rodriguez y Baena, R.; Gaetani, P.; Danova, M. Primary solitary intracranial melanoma: Case report and review of the literature. *Surg. Neurol.* **1992**, *38*, 26–37. [CrossRef]
56. Balakrishnan, R.; Porag, R.; Asif, D.S. Primary intracranial melanoma with early leptomeningeal spread: A case report and treatment options available. *Case Rep. Oncol. Med.* **2015**, *29*, 3802. [CrossRef]
57. Garbacz, T.; Osuchowski, M.; Bartosik-Psujek, H. Primary diffuse meningeal melanomatosis—A rare form of meningeal melanoma: Case report. *BMC Neurol.* **2019**, *19*, 271–275. [CrossRef] [PubMed]
58. Morais, S.; Cabral, A.; Santos, G.; Madeira, N. Melanoma brain metastates presenting as delirium: A case report. *Arch Clin. Psychiatry* **2017**, *44*, 53–54.
59. Lin, B.; Yang, H.; Qu, L.; Li, Y.; Yu, J. Primary meningeal melanocytoma of the anterior cranial fossa: A case report and review of the literature. *World J. Surg. Oncol.* **2012**, *10*, 135. [CrossRef] [PubMed]
60. Samuels, M.A. The brain-heart connection. *Circulation* **2007**, *116*, 77–84. [CrossRef] [PubMed]
61. Goldberg, E.M.; Schwartz, E.S.; Younkin, D.; Myers, S.R. Atypical Syncope in a child due to a colloid of the third ventricle. *Pediatr. Neurol.* **2011**, *45*, 331–334. [CrossRef] [PubMed]
62. Nikolaeov, S.I.; Rimoldi, D.; Iseli, C. Exome sequencing identifies recurrent somatic MAP2K1 and MAP2K2 mutations in melanoma. *Nat. Genet.* **2012**, *44*, 133–139. [CrossRef]

63. Stark, M.S.; Woods, S.L.; Gartside, M.G. Frequent somatic mutations in MAP3K5 and MAP3K9 in metastatic melanoma identified by exome sequencing. *Nat. Genet.* **2012**, *44*, 165–169. [CrossRef] [PubMed]
64. Kusters-Vandevelde, H.V.N.; Kusters, B.; Van Engen-Van Grusven, A.C.H.; Groenen, P.J.T.A.; Wesseling, P.; Blokx, W.A.M. Primary melanocytic tumors of the central nervous system: A review with focus on molecular aspects. *Brain Pathol.* **2015**, *25*, 209–226. [CrossRef] [PubMed]
65. Pan, Z.; Yang, G.; Wang, Y.; Yuan, T.; Gao, Y.; Dong, L. Leptomeningeal metastases from a primary central nervous system melanoma: A case report and literature review. *World J. Surg. Oncol.* **2014**, *12*, 265–272. [CrossRef]
66. Ma, Y.; Gui, Q.; Lang, S. Intracranial malignant melanoma: A report of 7 cases. *Oncol. Lett.* **2015**, *10*, 2171–2175. [CrossRef] [PubMed]
67. Isiklar, I.; Leeds, N.E.; Fuller, G.N.; Kumar, A.J. Intracranial metastatic melanoma: Correlation between MR imaging characteristics and melanin content. *Am. J. Roentgenol.* **1995**, *165*, 1503–1512. [CrossRef] [PubMed]
68. Chen, C.J.; Hsu, Y.I.; Ho, Y.S. Intracranial meningeal melanocytoma. CT and MRI. *Neuroradiology* **1997**, *39*, 811–814. [CrossRef] [PubMed]
69. Lee, H.J.; Ahn, B.C.; Hwang, S.W.; Cho, S.K.; Kim, H.W.; Lee, S.W.; Hwang, J.H.; Lee, J. F-18 fluorodeoxyglucose PET/CT and post hoc PET/MRI in a case of primary meningeal melanomatosis. *Korean J. Radiol.* **2013**, *14*, 343–349. [CrossRef] [PubMed]
70. Young, G.J.; Bi, W.L.; Wu, W.W.; Johanns, T.M.; Dung, G.P.; Dun, I.F. Management of intracranial melanomas in the era of precision medicine. *Oncotarget* **2017**, *8*, 89326–89347. [CrossRef]
71. Kircher, D.A.; Silvis, M.R.; Cho, J.H.; Holmen, S.L. Melanoma brain metastasis: Mechanism, models and medicine. *Int. J. Mol. Sci.* **2016**, *17*, 1468. [CrossRef]
72. Abate-Daga, D.; Ramello, M.C.; Smalley, I.; Forsyth, P.A.; Smalley, K.S.M. The biology and therapeutic management of melanoma brain metastases. *Biochem. Pharm.* **2018**, *153*, 35–45. [CrossRef]
73. Basnet, A.; Saad, N.; Benjamin, S. A case of vanishing brain metastasis in a melanoma patient on nivolumab. *Anticancer Res.* **2016**, *36*, 4795–4798. [CrossRef] [PubMed]
74. Puyana, C.; Denyer, S.; Burch, T.; Bhimani, A.D.; McGuire, L.S.; Patel, A.S.; Mehta, A.I. Primary Malignant Melanoma of the Brain: A Population-Based Study. *World Neurosurg.* **2019**, *130*, 1091–1097. [CrossRef] [PubMed]

 healthcare 

Article

Anaphylactic Death: A New Forensic Workflow for Diagnosis

Massimiliano Esposito [1,†], Angelo Montana [1,†], Aldo Liberto [1], Veronica Filetti [2], Nunzio Di Nunno [3], Francesco Amico [1], Monica Salerno [1,*], Carla Loreto [2] and Francesco Sessa [4]

1 Legal Medicine, Department of Medical, Surgical and Advanced Technologies, "G.F. Ingrassia", University of Catania, 95123 Catania, Italy; massimiliano.esposito91@gmail.com (M.E.); angelomontana49@gmail.com (A.M.); aldoliberto@gmail.com (A.L.); francescoamico08@gmail.com (F.A.)
2 Human Anatomy and Histology, Department of Biomedical and Biotechnology Sciences, University of Catania, 95123 Catania, Italy; verofiletti@gmail.com (V.F.); carla.loreto@unict.it (C.L.)
3 Department of History, Society and Studies on Humanity, University of Salento, 73100 Lecce, Italy; nunzio.dinunno@icloud.com
4 Department of Clinical and Experimental Medicine, University of Foggia, 71122 Foggia, Italy; francesco.sessa@unifg.it
* Correspondence: monica.salerno@unict.it; Tel.: +39-3735357201
† These authors contributed equally to this work.

Abstract: Anaphylaxis is a life-threatening or fatal clinical emergency characterized by rapid onset, and death may be sudden. The margin of certainty about the diagnosis of anaphylactic death is not well established. The application of immunohistochemical techniques combined with the evaluation of blood tryptase concentrations opened up a new field of investigation into anaphylactic death. The present study investigated eleven autopsy cases of anaphylactic death, carried out between 2005 and 2017, by the Departments of Forensic Pathology of the Universities of Foggia and Catania (Italy). An analysis of the medical records was carried out in all autopsies. Seven autopsies were carried out on males and four on females. Of the eleven cases, one showed a history of asthma, one of food ingestion, two of oral administration of medications, six did not refer any allergy history, and one subject was unknown. All cases (100%) showed pulmonary congestion and edema; 7/11 (64%) of the cases had pharyngeal/laryngeal edema and mucus plugging in the airway; only one case (9%) had a skin reaction that was found during external examination. Serum tryptase concentration was measured in ten cases, and the mean value was 133.5 μg/L ± 177.9. The immunohistochemical examination using an anti-tryptase antibody on samples from the lungs, pharynx/larynx, and skin site of medication injection showed that all cases (100%) were strongly immunopositive for anti-tryptase antibody staining on lung samples; three cases (30%) were strongly immunopositive for anti-tryptase antibody staining on pharyngeal/laryngeal samples; and eight cases (80%) were strongly immunopositive for anti-tryptase antibody staining on skin samples. We conclude that a typical clinical history, blood tryptase level >40 μg/L, and strongly positive anti-tryptase antibody staining in the immunohistochemical investigation may represent reliable parameters in the determination of anaphylactic death with the accuracy needed for forensic purposes.

Keywords: anaphylactic death; diagnostic workflow; immunohistochemical investigation; blood tryptase level

Citation: Esposito, M.; Montana, A.; Liberto, A.; Filetti, V.; Nunno, N.D.; Amico, F.; Salerno, M.; Loreto, C.; Sessa, F. Anaphylactic Death: A New Forensic Workflow for Diagnosis. *Healthcare* **2021**, *9*, 117. https://doi.org/10.3390/healthcare9020117

Academic Editor: Mariano Cingolani

Received: 18 December 2020
Accepted: 20 January 2021
Published: 22 January 2021

1. Introduction

The term anaphylaxis was introduced in 1902 by Portier and Richet [1], and it refers to a serious, generalized or systemic, allergic or hypersensitivity reaction [2]. It can be a life-threatening or fatal clinical emergency with airway and circulatory impairments [3–6]. It is usually associated with skin and mucosal alterations (widespread hives, pruritus, and swollen lips/tongue/uvula) and gastrointestinal disorders (vomiting, diarrhea, and abdominal cramps) [7,8]. In particular, anaphylaxis is due to a systemic reaction mediated by vasoactive amines, released from mast cells, and basophils sensitized by immunoglobulin E (IgE) [7,9–12]. Conversely, anaphylactic shock (AS) is an anaphylactic reaction characterized by critical organ hypoperfusion after exposure to a previously encountered antigen [11,12]. The incidence and prevalence of anaphylaxis are difficult to establish. However, the incidence ranges from 1.5 to 7.9 per 100,000 person-years, but there has been an increase in admissions with anaphylaxis over the last two decades [3]. Moreover, the prevalence is 0.3% in the European population [3]. According to Chaudhuri et al. [13], the incidence of anaphylaxis in the United States ranged from 1.21% to 15.04% in the population. Furthermore, the risk of severe anaphylaxis has been estimated to be 1–3 per 10,000 person-years, while the risk of death due to anaphylactic shock is about 1–3 per million per year. Food and medications are responsible for most anaphylaxis reactions. However, virtually any agent capable of directly or indirectly activating mast cells or basophils can cause this syndrome [9–18]. Food is the cause of anaphylaxis in children most of the time, and drugs are major causes in adults, and are also the most frequent cause of anaphylaxis in hospitalized patients [12,19–22]. A higher frequency of anaphylaxis has been shown in adult females to food and non-steroidal anti-inflammatory drugs (NSAIDs) [8,12]. The common drugs responsible for anaphylaxis reactions are antibiotics, muscle relaxants, non-steroidal anti-inflammatory drugs, and radioactive contrast media [8,13]. Risk factors for severe anaphylaxis with hospitalization are old age combined with comorbidities such as cardiovascular disease (CVD) and chronic obstructive pulmonary disease [14,15,22–26].

Tryptase is an abundant secretory granule-derived serine proteinase contained in mast cells. The tryptase enzyme is the only protein that is specific for human mast cells, and tryptase plasma levels reflect the clinical severity of anaphylaxis. Elevated levels of serum tryptase occur in both anaphylaxis and anaphylactic reactions, but a negative test does not exclude anaphylaxis [27].

In this study, an investigation of the serum tryptase levels combined with the immunohistochemical expression of tryptase in specimens from the lungs, glottis, and skin (site of medication injection) in eleven autopsy cases was performed to clarify and discuss their significance in anaphylactic death.

2. Materials and Methods

2.1. Sample Collection

A retrospective analysis of the autopsy records of the Departments of Forensic Pathology of the University of Foggia and Catania (Italy), was carried out between 2005 and 2017. From the analysis of death scene investigations and autopsy reports, together with the information gathered from the police, eleven cases of anaphylactic death origin were selected. Cases with weak or missing information about the manner of death were excluded. Decomposed bodies were also excluded from the study. All procedures performed in this study were in accordance with the ethical standards of the institution and with the 1964 Helsinki Declaration and its later amendments or comparable ethical standards. Informed consent was obtained from all relatives.

The deceased were four men and seven women, ranging in age from 16 to 69 years (average: 47 years $\pm$ 17.69). In all cases of anaphylactic death, sections of lungs, glottis and the skin site of medication injection were collected. Eleven cases were selected as controls. These included the following cases: seven cases of sudden cardiac death and four cases of fatal motor vehicle crashes. All control cases were selected for their negative clinical histories for manifestations of asthma or allergies. All autopsies were performed within four days after death was determined, and all cadavers were stored at -4 °C.

2.2. Histological Analysis

A routine microscopic histopathological study was performed using hematoxylin–eosin (H&E) staining. Specimens from the lungs, glottis and skin were fixed in 10% buffered formalin, as previously described [28]. After an overnight wash, specimens were dehydrated in graded ethanol, cleared in xylene and paraffin-embedded. Tissue paraffin blocks were then cut (4 μm thickness) using a microtome and sections were mounted on silane-coated slides (Dako, Glostrup, Denmark) and stored at room temperature. Sections then were stained with H&E and observed using a Zeiss Axioplan light microscope (Carl Zeiss, Oberkochen, Germany) for morphological examination. Finally, representative micrographs were captured using a Zeiss AxioCam MRc5 digital camera (Carl Zeiss, Oberkochen, Germany).

2.3. *Immunohistochemical Staining*

Immunohistochemical investigation of samples from the lungs, glottis and skin was performed using anti-tryptase antibodies. For the immunohistochemical analysis, specimens were processed as previously described [28,29]. In particular, sections were dewaxed in xylene, rehydrated with graded ethanol and then incubated for 20 min in 0.3% H_2O_2/methanol solution to block endogenous peroxidase activity. After rinsing for 20 min with phosphate buffered saline (PBS), slides were pre-treated to facilitate antigen retrieval and to increase membrane permeability to antibodies using a microwave oven (750 W) (5 min × 3) in capped polypropylene slide-holders with citrate buffer (10 mM citric acid, 0.05% Tween 20, pH 6.0; Bio-Optica, Milan, Italy) and then incubated overnight at 4 °C with anti-tryptase monoclonal antibodies (Agilent Dako, Copenhagen, Denmark) diluted 1:100 in PBS. The detection system used was the LSAB+ kit (Dako, Copenhagen, Denmark) incubated for 10 min at room temperature, a refined avidin–biotin technique in which a biotinylated secondary antibody reacts with several peroxidase-conjugated streptavidin molecules. The positive reaction was visualized by 3,3-diaminobenzidine (DAB) peroxidation (DAB substrate Chromogen System; Dako) according to standard methods [30]. The sections were counterstained with Mayer's hematoxylin (Histolab Products AB, Göteborg, Sweden) mounted in GVA (Zymed Laboratories, San Francisco, CA, USA). The sections were observed and photographed as described above.

2.4. *Evaluation of Immunohistochemistry (IHC)*

The anti-tryptase immunoreaction was identified as either negative or positive. Immunohistochemical positive staining was defined by the presence of brown chromogen detected on the edge of the hematoxylin-stained cell nucleus, distributed within the cytoplasm or in the membrane via evaluation by light microscopy. Positive controls consisted of tissue specimens with known antigenic positivity. Sections treated with PBS without any primary antibody served as negative controls. Seven fields of about 600,000 μm^2, randomly selected from each section, were considered for morphometric and densitometric analysis. The percentage of the areas (morphometric analysis) stained with anti-tryptase antibody was expressed as % positive dark brown pixels of the analyzed fields. Moreover, the levels (high/low) of staining intensity of positive areas (densitometric analysis) were expressed as densitometric count ($pixel^2$) of positive dark brown pixels in the analyzed fields. These parameters were calculated using software for image acquisition (AxioVision Release 4.8.2-SP2 Software, Carl Zeiss Microscopy GmbH, Jena, Germany). Data are expressed as mean ± standard deviation (SD). Digital micrographs were taken and fitted as previously described. The samples were also examined with a confocal microscope, and a three-dimensional reconstruction was performed (True Confocal Scanner, Leica TCS SPE).

2.5. *Statistical Analysis*

Statistical analysis was performed using GraphPad Prism 7.0 (GraphPad Software, Inc., La Jolla, CA, USA). The Shapiro–Wilk normality test was used for the calculation of the distribution of the samples. Unpaired t-tests were used for the comparison between the levels of tryptase staining intensity of positive areas ($pixel^2$) of cases of anaphylactic death and controls. p-values of less than 0.05 ($p < 0.05$) were considered significantly different.

2.6. Serum Tryptase Assay

Samples of femoral blood were obtained via a transcutaneous femoral approach (from the femoral artery) in eleven post-mortem examinations, of which all subsequently underwent full autopsy. Serum was derived from whole blood by centrifugation, decanted into plastic test-tubes and stored at −80 °C. Samples were shipped on ice to the Industrial Bio-Test (IBT) Reference Laboratory (Florence, Italy) for analysis. Information regarding the cause of death was hidden from the reference laboratory performing the assays. Serum tryptase levels were determined using a competitive immunofluorescent enzyme assay with monoclonal anti-human tryptase antibodies against both the A and B structural types of tryptase. These antibodies were incubated with a serum aliquot; the sample was washed, and enzyme-labelled anti-tryptase was added, followed by incubation. The sample was washed a second time, the developer was added, and the fluorescence in the aliquot was measured. The amount of fluorescence given off by the sample was directly proportional to the concentration of tryptase in the sample. Through a radioimmunoassay method, which only detected the β form of the tryptase molecule. Eleven cases were selected as controls. These included the following cases: seven cases of sudden cardiac death and four cases of death after motor vehicle crashes.

3. Results

Table 1 shows the clinical history of all selected cases, the cause of the anaphylactic reaction, and the interval between the onset of symptoms and death. According to the results shown in Table 1, 72.7% of our cases did not have a history of allergy; only 1/11 had a history of asthma and celiac disease; 5/11 (45.4%) died within 1 h, and 6/11 (54.6%) within 1 min. The causes of anaphylaxis were medications (6/11), injected contrast medium (3/11), food (1/11), and latex (1/11).

Table 1. The circumstantial data of the selected cases.

Case Report	History of Allergy	Time to Death	Cause
Case 1	Unknown	Within 1 h	Medications
Case 2	Unknown	Within 1 min	Medications
Case 3	Drugs	Within 1 min	Medications
Case 4	Unknown	Within 1 min	Contrast medium injected
Case 5	Asthma, celiac disease	Within 1 min	Food
Case 6	Unknown	Within 1 min	Contrast medium injected
Case 7	Food	Within 1 h	Latex
Case 8	Unknown	Within 1 h	Medications
Case 9	Unknown	Within 1 h	Medications
Case 10	Unknown	Within 1 h	Medications
Case 11	Unknown	Within 1 min	Contrast medium injected

The post-mortem diagnosis was based on: (1) the circumstantial evidence, including the history of exposure to a likely allergen prior to death and clinical presentation; (2) post-mortem findings suggesting an anaphylactic reaction, such as laryngeal edema, mucous plugging in the airways, erythematous skin rash and edema, eosinophilia in the mucosa and submucosa of the respiratory and the gastrointestinal tracts, and marked pulmonary congestion and edema; (3) toxicology test of serum concentrations of tryptase in femoral blood samples (from the femoral artery); (4) histological examination of all the organs using H&E; and (5) immunohistochemical examination for anti-tryptase antibody staining. The standard upper limit of total serum tryptase level has been established and was set at 40 μg/L in the Office of the Departments of Forensic Pathology.

The cases were analyzed regarding: (1) circumstantial evidence, including history of exposure; (2) post-mortem examination findings, including histological study and toxicology testing; and (3) cause of death. The data that were analyzed were extracted from the police investigation report, medical records, interviews of the victim's family members and reports by forensic pathologists or investigators, and forensic autopsy protocols.

On the basis of these results, a new workflow as a useful tool in anaphylaxis deaths was elaborated.

3.1. Autopsy Findings

All cases showed pulmonary swelling and edema during autopsy. Macroscopic examination during the autopsies revealed that 64% of the cases had pharyngeal/laryngeal edema and mucus plugging in the airways. Only one case (9%) had a skin reaction that was found during the external examination. The results of toxicological analyses performed on ante- and post-mortem samples (blood and urine) were negative for alcohol, drugs and medications.

3.2. Histological and Immunohistochemical Analysis

All cases displayed pulmonary congestion and edema during the histological examination. The glottis was sampled in six cases, and the skin was sampled in eight (only in cases of transdermal administration of medication). An immunohistochemical examination of anti-tryptase antibody staining on lung samples was performed in ten autopsies, glottis in six, and skin site of injected medications in eight. All cases showed strong immunopositivity for anti-tryptase antibody staining on lung samples (10/10), on pharyngeal/laryngeal samples (7/7), and on skin samples (8/8) (Table 2). Samples from the lung, skin site of injected medications and glottis showed a strong and diffuse anti-tryptase immunolabeling. In particular, in lung specimens, anti-tryptase was found in mast cells of the connective interstitium and bronchiolar structure (Figure 1a). The skin site of medication administration also showed strong mast cell antibody immunolabelling in the connective derma (Figure 1b). Moreover, the glottis of these cadavers exhibited an overexpression of anti-tryptase antibody staining scattered in the laminar connective tissue at the vocal fold level (Figure 1c). The Shapiro–Wilk normality test showed that the level of tryptase staining intensity of positive areas (pixel^2) in cases of anaphylactic death and controls differs significantly from a normal distribution. As shown in Figure 2, the level of tryptase staining intensity revealed that in the tissue of cadavers who had died from anaphylactic shock the tryptase immunostaining level was much higher compared to controls ($p < 0.05$). In particular, these results of tryptase immunostaining were confirmed in lung tissue (Figure 2a), skin tissue (Figure 2b), and glottis tissue (Figure 2c). Figure 3a summarizes the histological results (H&E), while in the other quadrants, the anti-tryptase immunohistochemical staining results are shown (Figure 3b–d). Figure 4 shows the anti-tryptase immunohistochemical results in lung samples by confocal laser scanning microscopy or with a light microscope (Figure 4a–d).

Table 2. Concentration of total serum tryptase, autopsy findings and histological examination. H&E: hematoxylin–eosin; IHC: immunohistochemistry.

Case Report	Tryptase Level in Blood (μg/L)	Autopsy Findings	Histological Examination
Case 1	136.5	Pulmonary congestion and edema	Pulmonary edema (H&E) Lung IHC not available
Case 2	130	Pulmonary congestion and edema Mucous plugging in the airways Glottis edema	Pulmonary edema (H&E) Lung + skin from medication injection site + glottis (IHC)
Case 3	200	Pulmonary congestion and edema Mucous plugging in the airways	Pulmonary edema (H&E) Lung + skin from medication injection site (IHC)
Case 4	640	Pulmonary congestion and edema Mucous plugging in the airways Glottis edema	Pulmonary edema (H&E) Lung (IHC)
Case 5	41.4	Pulmonary congestion and edema Mucous plugging in the airways	Pulmonary edema (H&E) Lung (IHC)
Case 6	290	Pulmonary congestion and edema Mucous plugging in the airways Glottis edema	Pulmonary edema (H&E) Lung + skin from medication injection site (IHC)
Case 7	133	Skin reaction Pulmonary congestion and edema Mucous plugging in the airways Glottis edema	Pulmonary edema (H&E) Lung + skin from medication injection site + glottis (IHC)
Case 8	160	Pulmonary congestion and edema Mucous plugging in the airways Glottis edema	Pulmonary edema (H&E) Lung + skin from medication injection site + glottis (IHC)
Case 9	40.5	Pulmonary congestion and edema Mucous plugging in the airways Glottis edema	Pulmonary edema (H&E) Lung + skin from medication injection site + glottis (IHC)
Case 10	83.6	Pulmonary congestion and edema Mucous plugging in the airways Glottis edema	Pulmonary edema (H&E) Lung + skin from medication injection site + glottis (IHC)
Case 11	89.6	Pulmonary congestion and edema Mucous plugging in the airways Pharyngeal/laryngeal edema	Pulmonary edema (H&E) Lung + skin from medication injection site + glottis (IHC)

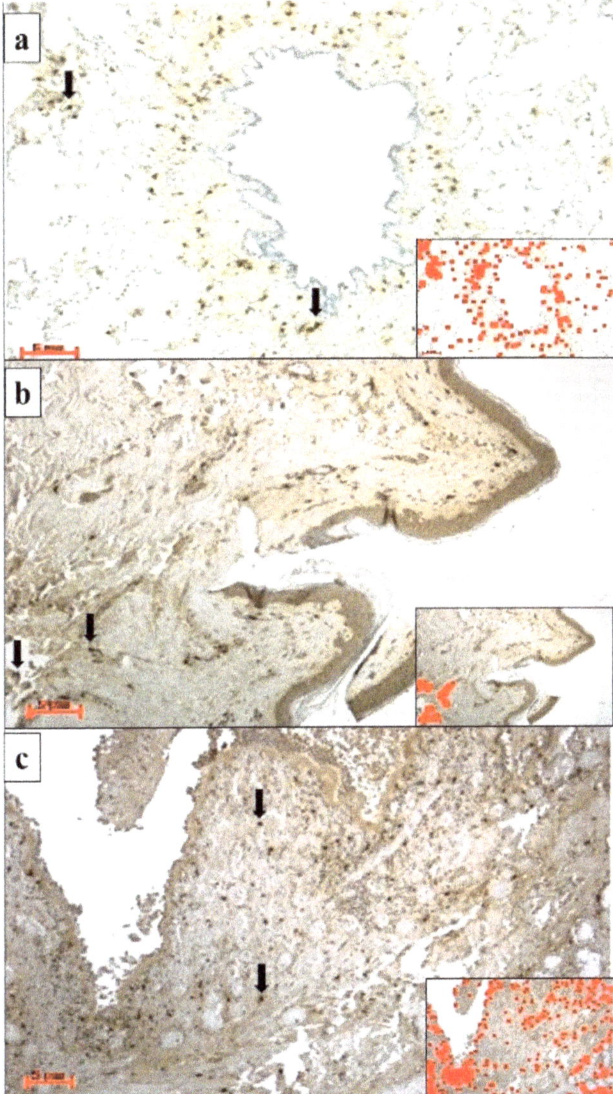

Figure 1. (**a**) Lung specimens from a cadaver who had died of anaphylactic shock; anti-tryptase antibody staining is strongly expressed in mast cells (black arrows) in the peribronchial interstitium. The insert shows the immunostaining software image analysis of Figure 1a, in which a highly immunostained area (red color) was detected (magnification: 20×; scale bar: 5 μm). (**b**) Skin specimens of the gluteus where medication administration occurred from a cadaver who had died of anaphylactic shock; anti-tryptase immunolocalization (black arrows) was demonstrated in the derma of the medication injection site. The insert shows the immunostaining software image analysis of Figure 1b, in which a highly immunostained area (red color) was detected (magnification: 20×; scale bar: 5 μm). (**c**) Glottis specimens of a cadaver who had died of anaphylactic shock showed strong anti-tryptase immunoexpression in mast cells (black arrows). The insert shows the immunostaining software image analysis of Figure 1c, in which a highly immunostained area (red color) was detected (magnification: 20×; scale bar: 5 μm).

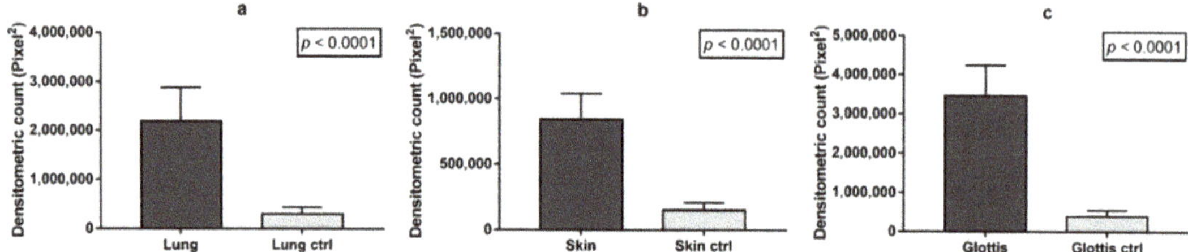

Figure 2. Comparison of the densitometric analysis (pixel^2) of the tryptase immunostained area expressed by positive, dark brown pixels in the analyzed fields for: (**a**) lung tissues (n = 10) of cadavers who had died of anaphylactic shock vs. controls (ctrl); (**b**) skin tissue (n = 8) of cadavers who had died of anaphylactic shock vs. ctrl; (**c**) glottis tissue (n = 7) of cadavers who had died of anaphylactic shock vs. ctrl. Data are presented as mean ± standard deviation (SD) ($p < 0.05$).

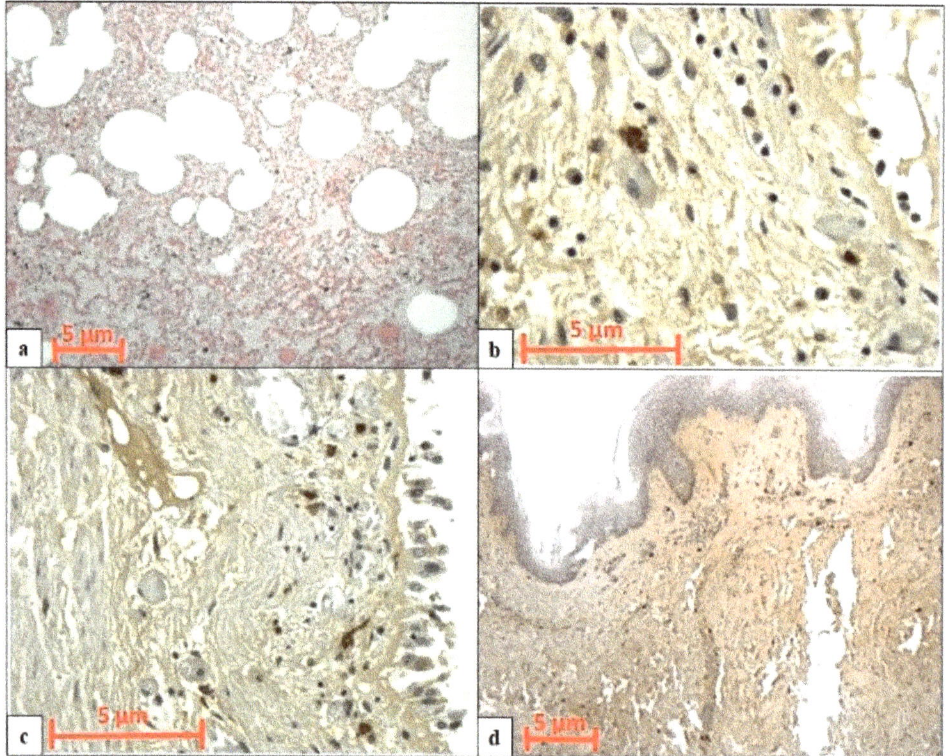

Figure 3. (**a**) H&E examination of lung samples shows capillary congestion and severe alveolar edema (amplification: 20×; scale bar: 5 µm). (**b**) IHC examination of lung samples with strong anti-tryptase immunopositivity (magnification: 40×; scale bar: 5 µm). (**c**) IHC examination of pharyngeal samples with strong anti-tryptase immunopositivity (magnification: 40×; scale bar: 5 µm). (**d**) IHC examination of skin samples with strong anti-tryptase immunopositivity (magnification: 20×; scale bar: 5 µm.

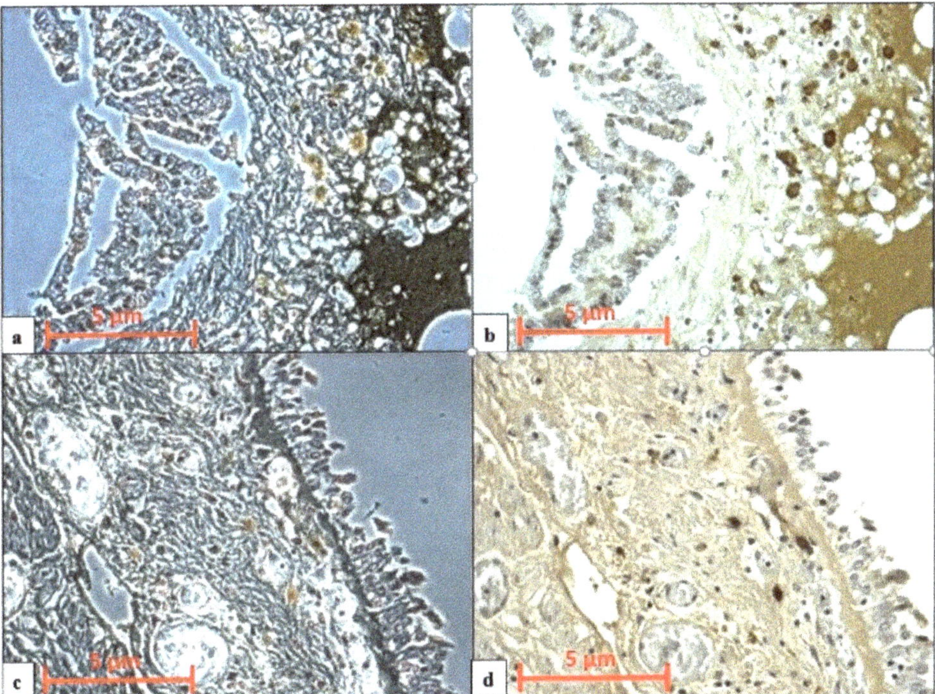

Figure 4. Comparison of anti-tryptase antibody reaction between confocal laser scanning microscopy (**a**,**c**) and light microscopy (**b**,**d**) shows strong anti-tryptase immunopositivity in: (**a**,**b**) lung tissue; (**c**,**d**) pharyngeal tissue. Magnification: 40×; scale bar: 5 µm.

3.3. *Serum Tryptase Analysis*

Routine toxicology tests were performed in all eleven autopsies. Serum tryptase analysis ranged from 40.5 µg/L to 640 µg/L, and the mean value was 133.5 µg/L ± 177.9.

Table 2 shows a summary of total serum tryptase, autopsy findings, and histological examinations. All cases showed pulmonary congestion and edema during autopsy and the histological examination.

4. Discussion

Today, there is no specific forensic workflow in cases of death from anaphylactic shock. A systematic approach would allow forensic pathologists to arrive at a confident diagnosis of death from anaphylactic shock. Through a retrospective analysis of eleven deaths from anaphylactic shock, the aim of this study was to propose a new forensic workflow, leading to a more accurate diagnosis.

The present study shows that a typical clinical history, high levels of serum tryptase (>40 µg/L), and strong positivity for anti-tryptase antibody staining are highly suggestive for establishing the diagnosis of anaphylactic death.

Anaphylaxis is a life-threatening syndrome [3–6]. The anaphylactic reaction is mostly triggered by food and drugs, but it may be provoked by any agent capable of activating mast cells or basophils [9,24]. In the UK, about half of the 20 fatal reactions recorded each year due to anaphylactic shock are iatrogenic; the rest are caused by food ingestion or insect venom. Respiratory or cardiac arrest occurs within 30 min for food, 15 min for venom, and 5 min for iatrogenic reactions [30]. A history of exposure to anaphylactic stimuli and clinical features such as hypotension are important to identify death from anaphylactic shock [31].

Several epidemiologic and experimental studies [9,28,32–46] have underlined the importance of immunohistochemical analyses and the concentrations of serum tryptase; however, based on the literature, this article is the first study that combines the two parameters for a specific diagnosis of anaphylactic death.

This study showed that the symptoms of the anaphylactic reaction occurred within one hour (Table 1): one minute in the case of injected contrast medium reaction, in cases of anaphylaxis during anesthesia, the shock occurred within one minute; in cases of medication, anaphylaxis shock occurred both within one minute (60%) and within one hour (40%). These results are in agreement with previous studies [3–6].

The autopsy procedure has to be careful, with an accurate external examination, searching for injection sites of stinging or biting invertebrates, as well as blood, vitreous, and urine collection. It is essential to examine the stomach contents, above all in suspected cases of anaphylactic shock from food [47]. Autopsy findings, such as the formation of mucus plugs, congestion and intra alveolar hemorrhages, and congestion and edema of major organs, are not exhaustive or specific for the diagnosis of fatal anaphylaxis [37]. Immunohistochemical analysis using anti-tryptase antibodies is also not exhaustive for the diagnosis of death from anaphylactic shock [48–50].

According to Turillazzi et al. [39], in all cases, the larynx and pharynx were opened with forceps following the posterior median line and the glottis was observed; the sides were stretched outward to study the mucosa. Then, following the Ghon technique, abdominal organs were removed using the bloc method, taking care to preserve the integrity of vascular structures. All autopsies showed pulmonary congestion and edema of the lungs. When squeezing the lungs, in all cases, an abundant, reddish-colored liquid was observed. Macroscopic examination during the autopsies showed glottis edema and mucus plugging in the airways in 64% of cases. Only one case had a skin reaction that was found during the external examination.

In our study, the results of the histological and immunohistochemical analyses showed generalized stasis with areas of acute pulmonary emphysema in all autopsies, and in deaths of subjects over 40 years old, eosinophilic cross-bands ranging from segments of hypercontracted to coagulated sarcomeres in heart samples. The immunohistochemical examination of anti-tryptase antibody staining on samples from the lungs, glottis, and skin of medication injection sites revealed strong positivity for anti-tryptase antibody staining for all sampling sites in all cases. In particular, in lung specimens, anti-tryptase was found in mast cells of the connective interstitium and bronchiolar structure. Skin sites of medication administration also showed strong mast cell antibody immunolabelling in the connective derma. Moreover, the glottis of these cadavers exhibited a high level of anti-tryptase antibody staining scattered in the laminar connective tissue at the vocal fold level.

According to the guidelines on autopsy practice for suspected acute anaphylaxis of the Royal College of Pathologists [49], serum tryptase samples should always be collected, even if an autopsy is performed days or even weeks after death. Despite an average serum tryptase concentration, anaphylactic death cannot be completely excluded. Different sampling techniques can impact post-mortem tryptase levels [49,51]. A recent study demonstrated that the level of tryptase is significantly lower in samples collected via transcutaneous aspiration compared with femoral/external iliac vein samples [38]. In fact, for post-mortem tryptase analysis, a sample from a clamped femoral/external iliac vein should be defined as the gold standard [38,52,53]. There are doubts about the variability of serum tryptase by post-mortem interval (PMI). Mast cells present in the respiratory tract and heart and post-mortem cell lysis might influence the release of tryptase; for this reason, peripheral blood (i.e., femoral blood) is preferable to central blood [49,53]. After death, mast cell tryptase is very stable with a long half-life, and it could be measured up to four days after death [49].

There are four different tryptases (α, β, γ, and δ), but only the α and β form are medically necessary; during an anaphylaxis reaction, they are released by mast cells [10]. Tryptase has proinflammatory effects such as the promotion of tissue edema and remodeling, chemokine secretion, and neutrophil recruitment [10]. The tryptase level can be increased by cell autolysis or liquefaction [54]. Higher values of tryptase serum have also been found in other types of death, such as sudden infant death syndrome (SIDS), amniotic fluid embolism, and heroin-related deaths [53–59]. β-tryptase is a more reliable indicator of acute mast cell activation. It is emitted at the same time as histamine, but the release is slower, making this marker more suitable for post-mortem investigation [60]. In 1998, Edison et al. [51] proposed that the cut-off level of tryptase should be 10 μg/L. Subsequently, in 2007, Edston et al. [45] modified the value to over 20 (44.5 μg/L) in femoral samples. In 2011, Mayer et al. [44] recommended a cut-off level over 45 μg/L. In 2014, McLean-Tooke et al. [30] modified the cut-off level to 110 μg/L in aortic samples. Finally, in 2017, Xiao et al. [48] established a cut-off level of 43 μg/L using femoral samples. There are few studies on the change in the cut-off of tryptase levels in cases of cardiac blood samples. However, it is easy to consider that the cut-off is the same as for aortic sampling (110 μg/L) with the same reliable margin (sensitivity of 80% and specificity of 92.1%). According to Tse et al. [42], sensitivity reaches 100% when the cut-off of tryptase is between 11.4 and 30 (μg/L), but specificity is low; specificity reaches 100% when the cut-off is above 70 (μg/L). In the case of aortic blood samples, the cut-off is 110 μg/L (sensitivity 80% and specificity 92.1%) [30,37].

In this retrospective analysis, tryptase serum determination was performed as part of all autopsies (11/11). The concentration ranged from 40.5 μg/L to 640 μg/L with a median of 133.5 μg/L ± 177.9. All autopsies were performed within four days after death and all cadavers were stored at −4 °C; this did not change the validity of the test. In fact, storing a corpse at −4 °C after death does not affect tryptase levels, as has been shown by previous studies. Sravan et al. [50] performed autopsies three days after death with storage at 4 °C. Edston et al. [45] published their study in which the mean time between death and autopsy was 3.861days. Tse [42] reported two cases in which there was an analysis of tryptase levels at three days and six days after death.

The combination of anamnestic information, autopsy findings, tryptase serum determination, and immunohistochemical testing can help to make a diagnosis of anaphylactic reaction as the cause of death in patients who died suddenly with unspecific symptoms.

The post-mortem diagnosis of anaphylactic shock is a challenge, and it is often achieved by exclusion. A limitation of this study is the small sample size of the analysis. In this regard, we suggest future studies to confirm our observations.

A sampling of serum tryptase is mandatory [49]. However, a literature review revealed that there are many doubts about its cut-off, sampling site (central or peripheral), and changes during the post-mortem interval (PMI). Histological and immunohistochemical investigation, through the use of the confocal microscope, help in the diagnosis. The results of the present study suggest that through the use of the blood tryptase concentration, together with the immunohistochemical investigation for anti-tryptase antibody staining in samples from the lung, glottis, and skin (at the site of administration of medications and contrast medium), it is possible to realize a very reliable diagnostic workflow of anaphylactic death (Figure 5). In fact, previous studies reported in the literature [61–68] have not clearly expressed how to establish a specific diagnosis of anaphylactic death. This diagnostic workflow should be used to establish an anaphylactic reaction as the cause of death with a large margin of certainty.

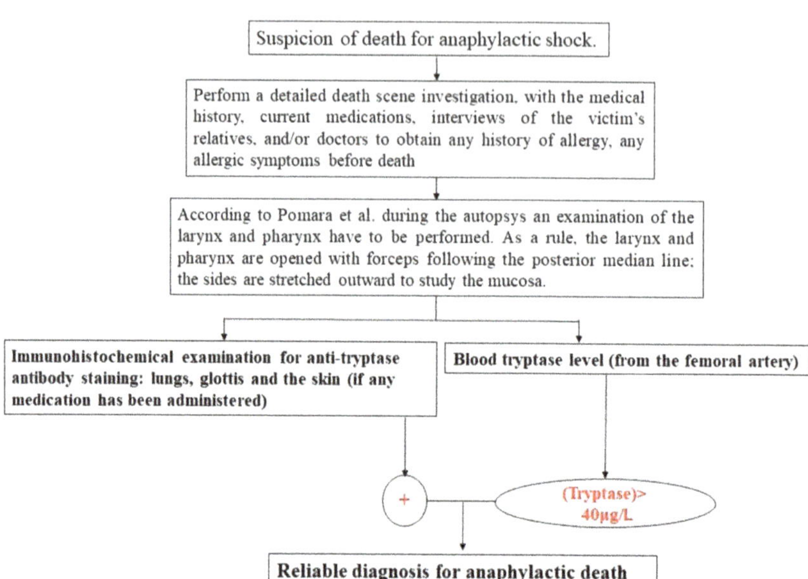

Figure 5. Proposed workflow to detect anaphylactic shock in fatal cases. In red text the goal of the flow chart.

Author Contributions: Conceptualization, M.E. and A.M.; methodology, V.F. and C.L.; software, C.L.; validation, N.D.N., A.L. and M.S.; formal analysis, V.F.; investigation, C.L.; resources, M.E.; data curation, F.A.; writing—original draft preparation, M.E.; writing—review and editing, M.E.; visualization, M.S.; supervision, C.L. and F.S. All authors have read and agreed to the published version of the manuscript.

Funding: This research received no external funding.

Institutional Review Board Statement: All procedures performed in the study were approved by the Scientific Committee of the University of Foggia, (code: 09/2018) and were performed in accordance with the 1964 Helsinki Declaration and its later amendments or comparable ethical standards.

Informed Consent Statement: Informed consent was obtained from all subjects involved in the study.

Data Availability Statement: All data are included in the main text.

Acknowledgments: We wish to thank the Scientific Bureau of the University of Catania for language support.

Ethical Approval and Consent to Participate: All procedures performed in the study were in accordance with the ethical standards of the institution and with the 1964 Helsinki Declaration and its later amendments or comparable ethical standards. Informed consent was obtained from the relatives.

Conflicts of Interest: The authors declare no conflict of interest.

References

1. Portier, M.M.; Richet, C. De l'action anaphylactique de certains venins. Comptesrendus hebdomadaires des séances et mémoires de la société de Biologie, séance du 15 février. *CR Soc. Biol.* **1902**, *54*, 170.
2. Simons, F.E.R.; Ardusso, L.R.; Bilò, M.B.; Cardona, V.; Ebisawa, M.; El-Gamal, Y.M.; Sanchez-Borges, M. International consensus on (ICON) anaphylaxis. *World Allergy Organ.* **2014**, *7*, 9. [CrossRef] [PubMed]
3. Muraro, A.; Roberts, G.; Worm, M.; Bilò, M.B.; Brockow, K.; Fernández Rivas, M.; Bindslev-Jensen, C. Anaphylaxis: Guidelines from the European Academy of Allergy and Clinical Immunology. *Allergy* **2014**, *69*, 1026–1045. [CrossRef]
4. Simons, F.E.R.; Ardusso, L.R.; Bilò, M.B.; El-Gamal, Y.M.; Ledford, D.K.; Ring, J.; Thong, B.Y. World Allergy Organization anaphylaxis guidelines: Summary. *J. Allergy Clin. Immunol.* **2011**, *127*, 587–593. [CrossRef] [PubMed]

5. Johansson, S.G.O.; Bieber, T.; Dahl, R.; Friedmann, P.S.; Lanier, B.Q.; Lockey, R.F.; Thien, F. Revised nomenclature for allergy for global use: Report of the Nomenclature Review Committee of the World Allergy Organization, October 2003. *J. Allergy Clin. Immunol.* **2004**, *113*, 832–836. [CrossRef]
6. Pumphrey, R.S.H. Lessons for management of anaphylaxis from a study of fatal reactions. *Clin. Exp. Aallergy* **2000**, *30*, 1144–1150. [CrossRef]
7. Sampson, H.A.; Muñoz-Furlong, A.; Campbell, R.L.; Adkinson, N.F., Jr.; Bock, S.A.; Branum, A.; Gidudu, J. Second symposium on the definition and management of anaphylaxis: Summary report—Second National Institute of Allergy and Infectious Disease/Food Allergy and Anaphylaxis Network symposium. *J. Allergy Clin. Immunol.* **2006**, *117*, 391–397. [CrossRef]
8. Moneret-Vautrin, D.A.; Morisset, M.; Flabbee, J.; Beaudouin, E.; Kanny, G. Epidemiology of life-threatening and lethal anaphylaxis: A review. *Allergy* **2005**, *60*, 443–451. [CrossRef]
9. Kemp, S.F.; Lockey, R.F. Anaphylaxis: A review of causes and mechanisms. *J. Allergy Clin. Immunol.* **2002**, *110*, 341–348. [CrossRef]
10. Reber, L.L.; Hernandez, J.D.; Galli, S.J. The pathophysiology of anaphylaxis. *J. Allergy Clin. Immunol.* **2017**, *140*, 335–348. [CrossRef]
11. Jeppesen, A.N.; Christiansen, C.F.; Frøslev, T.; Sørensen, H.T. Hospitalization rates and prognosis of patients with anaphylactic shock in Denmark from 1995 through 2012. *J. Allergy Clin. Immunol.* **2016**, *137*, 1143–1147. [CrossRef] [PubMed]
12. González-Pérez, A.; Aponte, Z.; Vidaurre, C.F.; Rodríguez, L.A.G. Anaphylaxis epidemiology in patients with and patients without asthma: A United Kingdom database review. *J. Allergy Clin. Immunol.* **2010**, *125*, 1098–1104. [CrossRef] [PubMed]
13. Chaudhuri, K.; Gonzales, J.; Jesurun, C.A.; Ambat, M.T.; Mandal-Chaudhuri, S. Anaphylactic shock in pregnancy: A case study and review of the literature. *Int. J. Obs. Anesth.* **2008**, *17*, 350–357. [CrossRef] [PubMed]
14. Neugut, A.I.; Ghatak, A.T.; Miller, R.L. Anaphylaxis in the United States: An investigation into its epidemiology. *Arch. Intern. Med.* **2001**, *161*, 15–21. [CrossRef] [PubMed]
15. Lieberman, P.; Simons, F.E.R. Anaphylaxis and cardiovascular disease: Therapeutic dilemmas. *Clin. Exp. Allergy* **2015**, *45*, 1288–1295. [CrossRef]
16. Fauci, A.S.; Kasper, D.L.; Hauser, S.L.; Jameson, J.L.; Loscalzo, J. *Harrison's Principles of Internal Medicine*; Longo, D.L., Ed.; Mcgraw-Hill: New York, NY, USA, 2012.
17. Lieberman, P.L. Specific and idiopathic anaphylaxis: Pathophysiology and treatment. In *Allergy, Asthma and Immunology from Infancy to Adulthood*, 3rd ed.; WB Saunders: Philadelphia, PA, USA, 1996; pp. 297–319.
18. Kaliner, M.; Sigler, R.; Summers, R.; Shelhamer, J.H. Effects of infused histamine: Analysis of the effects of H-1 and H-2 histamine receptor antagonists on cardiovascular and pulmonary responses. *J. Allergy Clin. Immunol.* **1981**, *68*, 365–371. [CrossRef]
19. Mitsuhata, H.; Shimizu, R.; Yokoyama, M.M. Role of nitric oxide in anaphylactic shock. *J. Clin. Immunol.* **1995**, *15*, 277–283. [CrossRef]
20. Lin, R.Y.; Schwartz, L.B.; Curry, A.; Pesola, G.R.; Knight, R.J.; Lee, H.S.; Westfal, R.E. Histamine and tryptase levels in patients with acute allergic reactions: An emergency department–based study. *J. Allergy Clin. Immunol.* **2000**, *106*, 65–71. [CrossRef]
21. Raper, R.F.; Fisher, M.M. Profound reversible myocardial depression after anaphylaxis. *Lancet* **1988**, *331*, 386–388. [CrossRef]
22. Kovanen, P.T.; Kaartinen, M.; Paavonen, T. Infiltrates of activated mast cells at the site of coronary atheromatous erosion or rupture in myocardial infarction. *Circulation* **1995**, *92*, 1084–1088. [CrossRef]
23. Kemp, S.F.; Lockey, R.F. Mechanisms of Anaphylaxis. In *Allergy Frontiers: Clinical Manifestations*; Springer: Tokyo, Japan, 2009; pp. 367–377.
24. Mehr, S.S.; Tey, D.; Tang, M.L.K. Paediatric anaphylaxis: A 5 year retrospective review. *Allergy* **2008**, *63*, 1071–1076.
25. Worm, M.; Edenharter, G.; Rueff, F.; Scherer, K.; Pföhler, C.; Mahler, V.; Niggemann, B. Symptom profile and risk factors of anaphylaxis in Central Europe. *Allergy* **2012**, *67*, 691–698. [CrossRef] [PubMed]
26. Vetander, M.; Helander, D.; Flodström, C.; Östblom, E.; Alfven, T.; Ly, D.H.; Wickman, M. Anaphylaxis and reaction to foods in children–a population-based case study of emergency department visits. *Clin. Exp. Allergy* **2012**, *42*, 568–577. [CrossRef] [PubMed]
27. Sun, K.J.; He, J.T.; Huang, H.Y.; Xue, Y.; Xie, X.L.; Wang, Q. Diagnostic role of serum tryptase in anaphylactic deaths in forensic medicine: A systematic review and meta-analysis. *Forensic Sci. Med. Pathol.* **2018**, *14*, 209–215. [CrossRef] [PubMed]
28. Castorina, S.; Lombardo, C.; Castrogiovanni, P.; Musumeci, G.; Barbato, E.; Almeida, L.E.; Leonardi, R. P53 and VEGF expression in human temporomandibular joint discs with derangement correlate with degeneration. *J. Biol. Regul. Homeost Agents* **2019**, *33*, 1657–1662.
29. Loreto, C.; Filetti, V.; Almeida, L.E.; La Rosa, G.R.M.; Leonardi, R.; Grippaudo, C.; Lo Giudice, A. MMP-7 and MMP-9 are overexpressed in the synovial tissue from severe temporomandibular joint dysfunction. *Eur. J. Histochem.* **2020**, *64*, 3113. [CrossRef]
30. McLean-Tooke, A.; Goulding, M.; Bundell, C.; White, J.; Hollingsworth, P. Postmortem serum tryptase levels in anaphylactic and non-anaphylactic deaths. *J. Clin. Pathol.* **2014**, *67*, 134–138. [CrossRef]
31. Leonardi, R.; Talic, N.F.; Loreto, C. MMP-13 (collagenase 3) immunolocalisation during initial orthodontic tooth movement in rats. *Acta Histochem.* **2007**, *109*, 215–220. [CrossRef]
32. Nara, A.; Aki, T.; Funakoshi, T.; Uchida, K.; Nakayama, H.; Uemura, K. Death due to blood transfusion-induced anaphylactic shock: A case report. *Leg. Med.* **2010**, *12*, 148–150. [CrossRef]

33. Yilmaz, R.; Yuksekbas, O.; Erkol, Z.; Bulut, E.R.; Arslan, M.N. Postmortem findings after anaphylactic reactions to drugs in Turkey. *Am. J. Forensic Med. Pathol.* **2009**, *30*, 346–349. [CrossRef]
34. Pumphrey, R.S.; Roberts, I.S. Postmortem findings after fatal anaphylactic reactions. *J. Clin. Pathol.* **2000**, *53*, 273–276. [CrossRef] [PubMed]
35. Shen, Y.; Li, L.; Grant, J.; Rubio, A.; Zhao, Z.; Zhang, X.; Fowler, D. Anaphylactic deaths in Maryland (United States) and Shanghai (China): A review of forensic autopsy cases from 2004 to 2006. *Forensic Sci. Int.* **2009**, *186*, 1–5. [CrossRef] [PubMed]
36. Unkrig, S.; Hagemeier, L.; Madea, B. Postmortem diagnostics of assumed food anaphylaxis in an unexpected death. *Forensic Sci. Int.* **2010**, *198*, e1–e4. [CrossRef]
37. Cecchi, R. Diagnosis of anaphylactic death in forensics: Review and future perspectives. *Leg. Med.* **2016**, *22*, 75–81. [CrossRef] [PubMed]
38. Platzgummer, S.; Bizzaro, N.; Bilò, M.B.; Pravettoni, V.; Cecchi, L.; Sargentini, V.; Bagnasco, M. Recommendations for the use of tryptase in the diagnosis of anaphylaxis and clonal mastcell disorders. *Allergy Clinimmunol.* **2020**, *52*, 51–61. [CrossRef]
39. Turillazzi, E.; Pomara, C.; La Rocca, G.; Neri, M.; Riezzo, I.; Karch, S.B.; Fineschi, V. Immunohistochemical Marker for Na+ CP Type Vα (C-20) and Heterozygous Nonsense SCN5A Mutation W822X in a Sudden Cardiac Death Induced by Mild Anaphylactic Reaction. *Appl. Immunohistochem. Mol. Morphol.* **2009**, *17*, 357–362. [CrossRef]
40. Hitosugi, M.; Omura, K.; Yokoyama, T.; Kawato, H.; Motozawa, Y.; Nagai, T.; Tokudome, S. 2. An autopsy case of fatal anaphylactic shock following fluorescein angiography. *Med. Sci. Law* **2004**, *44*, 264–265. [CrossRef]
41. Turillazzi, E.; Greco, P.; Neri, M.; Pomara, C.; Riezzo, I.; Fineschi, V. Anaphylactic latex reaction during anaesthesia: The silent culprit in a fatal case. *Forensic Sci. Int.* **2008**, *179*, e5–e8. [CrossRef]
42. Tse, R.; Garland, J.; Ahn, Y. Decline in 2 serial postmortem tryptase measurements beyond 72 hours after death in an antibiotic-related anaphylactic death. *Am. J. Forensic Med. Pathol.* **2018**, *39*, 14–17. [CrossRef]
43. Comment, L.; Bonetti, L.R.; Mangin, P.; Palmiere, C. Measurement of β-tryptase in postmortem serum, pericardial fluid, urine and vitreous humor in the forensic setting. *Forensic Sci. Int.* **2014**, *240*, 29–34. [CrossRef]
44. Mayer, D.E.; Krauskopf, A.; Hemmer, W.; Moritz, K.; Jarisch, R.; Reiter, C. Usefulness of post mortem determination of serum tryptase, histamine and diamine oxidase in the diagnosis of fatal anaphylaxis. *Forensic Sci. Int.* **2011**, *212*, 96–101. [CrossRef] [PubMed]
45. Edston, E.; Eriksson, O.; van Hage, M. Mast Cell Tryptase Postmortem Serum—Ref. Values Confounders. *Int. J. Leg. Med.* **2007**, *121*, 275–280. [CrossRef] [PubMed]
46. Garland, J.; Philcox, W.; McCarthy, S.; Mathew, S.; Hensby-Bennett, S.; Ondrushka, B.; Ahn, Y. Postmortem Tryptase Level in 120 Consecutive Nonanaphylactic Deaths: Establishing a Reference Range as <23 μg/L. *Am. J. Forensic Med. Pathol.* **2019**, *40*, 351–355. [PubMed]
47. Garland, J.; Philcox, W.; McCarthy, S.; Hensby-Bennet, S.; Ondruschka, B.; Woydt, L.; Kesha, K. The effects of different sampling techniques on peripheral post mortem tryptase levels: A recommended sampling method. *Int. J. Leg. Med.* **2019**, *133*, 1477–1483. [CrossRef]
48. Xiao, N.; Li, D.R.; Wang, Q.; Zhang, F.; Yu, Y.G.; Wang, H.J. Postmortem serum tryptase levels with special regard to acute cardiac deaths. *J. Forensic Sci.* **2017**, *62*, 1336–1338. [CrossRef]
49. Wilkins, B. Guidelines on Autopsy Practice: Autopsy for Suspected Acute Anaphylaxis (Includes Anaphylactic Shock and Anaphylactic Asthma). 2018. Available online: https://www.rcpath.org/uploads/assets/47841b6b-891f-450a-b968889ff3e0a7d1/G170-DRAFT-Guidelines-on-autopsy-practice-autopsy-for-suspected-acute-anaphalaxis-For-Consultation.pdf (accessed on 20 January 2021).
50. Sravan, A.; Tse, R.; Cala, A.D. A decline in 2 consecutive postmortem serum tryptase levels in an anaphylactic death. *Am. J. Forensic Med. Pathol.* **2015**, *36*, 233–235. [CrossRef]
51. Edston, E.; van Hage-Hamsten, M. β-Tryptase measurements post-mortem in anaphylactic deaths and in controls. *Forensic Sci. Int.* **1998**, *93*, 135–142. [CrossRef]
52. Edston, E. Accumulation of eosinophils, mast cells, and basophils in the spleen in anaphylactic deaths. *Forensic Sci. Med. Pathol.* **2013**, *9*, 496–500. [CrossRef]
53. Patanè, F.G.; Liberto, A.; Maglitto, A.N.M.; Malandrino, P.; Esposito, M.; Amico, F.; Cocimano, G. Nandrolone Decanoate: Use, Abuse and Side Effects. *Medicina* **2020**, *56*, 606. [CrossRef]
54. Nishio, H.; Takai, S.; Miyazaki, M.; Horiuchi, H.; Osawa, M.; Uemura, K.; Suzuki, K. Usefulness of serum mast cell–specific chymase levels for postmortem diagnosis of anaphylaxis. *Int. J. Leg. Med.* **2005**, *119*, 331–334. [CrossRef]
55. Neri, M.; Fabbri, M.; D'Errico, S.; Di Paolo, M.; Frati, P.; Gaudio, R.M.; Turillazzi, E. Regulation of miRNAs as new tool for cutaneous vitality lesions demonstration in ligature marks in deaths by hanging. *Sci. Rep.* **2019**, *9*, 20011. [CrossRef]
56. Riches, K.J.; Gillis, D.; James, R.A. An autopsy approach to bee sting-related deaths. *Pathology* **2002**, *34*, 257–262. [CrossRef]
57. Albano, G.D.; Bertozzi, G.; Maglietta, F.; Montana, A.; Di Mizio, G.; Esposito, M.; Salerno, M. Medical Records Quality as Prevention Tool for Healthcare-Associated Infections (HAIs) Related Litigation: A Case Series. *Curr. Pharm. Biotechnol.* **2019**, *20*, 653–657. [CrossRef]
58. Burkhardt, S.; Genet, P.; Sabatasso, S.; La Harpe, R. Death by anaphylactic shock in an institution: An accident or negligence? *Int. J. Leg. Med.* **2019**, *133*, 561–564. [CrossRef]

59. Turillazzi, E.; Vacchiano, G.; Luna Maldonado, A.; Neri, M.; Pomara, C.; Rabozzi, R.; Fineschi, V. Tryptase, CD15 and IL-15 as reliable markers for the determination of soft and hard ligature marks vitality. *Histol. Histopathol.* **2010**, *25*, 1539–1546.
60. Fineschi, V.; Cecchi, R.; Centini, F.; Reattelli, L.P.; Turillazzi, E. Immunohistochemical quantification of pulmonary mast-cells and post-mortem blood dosages of tryptase and eosinophil cationic protein in 48 heroin-related deaths. *Forensic Sciint* **2001**, *120*, 189–194. [CrossRef]
61. Pomara, C.; Fineschi, V. *Forensic and Clinical Forensic Autopsy: An Atlas and Handbook*; CRC Press: Boca Raton, FL, USA, 2020.
62. Torrisi, M.; Pennisi, G.; Russo, I.; Amico, F.; Esposito, M.; Liberto, A.; Cocimano, G. Sudden Cardiac Death in Anabolic-Androgenic Steroid Users: A Literature Review. *Medicina* **2020**, *56*, 587. [CrossRef]
63. Montana, A.; Salerno, M.; Feola, A.; Asmundo, A.; Di Nunno, N.; Casella, F.; Manno, E.; Colosimo, F.; Serra, R.; Di Mizio, G. Risk Management and Recommendations for the Prevention of Fatal Foreign Body Aspiration: Four Cases Aged 1.5 to 3 Years and Mini-Review of the Literature. *Int. J. Environ. Res. Public Health* **2020**, *17*, 4700. [CrossRef]
64. Strisciuglio, T.; Di Gioia, G.; De Biase, C.; Esposito, M.; Franco, D.; Trimarco, B.; Barbato, E. Genetically determined platelet reactivity and related clinical implications. *High Blood Press. Cardiovasc. Prev.* **2015**, *22*, 257–264. [CrossRef]
65. Peters, F.T.; Wissenbach, D.K.; Busardo, F.P.; Marchei, E.; Pichini, S. Method development in forensic toxicology. *Curr. Pharm. Des.* **2017**, *23*, 5455–5467. [CrossRef]
66. Busardò, F.P.; Frati, P.; Zaami, S.; Fineschi, V. Amniotic fluid embolism pathophysiology suggests the new diagnostic armamentarium: β-tryptase and complement fractions C3-C4 are the indispensable working tools. *Int. J. Mol. Sci.* **2015**, *16*, 6557–6570. [CrossRef] [PubMed]
67. Busardò, F.P.; Marinelli, E.; Zaami, S. Is the diagnosis of anaphylaxis reliable in forensics? The role of β-tryptase and its correct interpretation. *Leg. Med.* **2016**, *23*, 86.
68. Zaami, S.; Busardò, F.P. Total tryptase or β-tryptase in post mortem settings: Which is to be preferred? *Forensic Sci. Int.* **2018**, *290*, e34. [CrossRef] [PubMed]

Communication

Microbiome Forensic Biobanking: A Step toward Microbial Profiling for Forensic Human Identification

Luciana Caenazzo and **Pamela Tozzo** *

Laboratory of Forensic Genetics, Department of Molecular Medicine, University of Padova, 35121 Padova, Italy; luciana.caenazzo@unipd.it

* Correspondence: pamela.tozzo@unipd.it

Abstract: In recent years many studies have highlighted the great potential of microbial analysis in human identification for forensic purposes, with important differences in microbial community composition and function across different people and locations, showing a certain degree of uncertainty. Therefore, further studies are necessary to enable forensic scientists to evaluate the risk of microbial transfer and recovery from various items and to further critically evaluate the suitability of current human DNA recovery protocols for human microbial profiling for identification purposes. While the establishment and development of microbiome research biobanks for clinical applications is already very structured, the development of studies on the applicability of microbiome biobanks for forensic purposes is still in its infancy. The creation of large population microbiome biobanks, specifically dedicated to forensic human identification, could be worthwhile. This could also be useful to increase the practical applications of forensic microbiology for identification purposes, given that this type of evidence is currently absent from most real casework investigations and judicial proceedings in courts.

Keywords: microbiome; human identification; biobank; forensic

Citation: Caenazzo, L.; Tozzo, P. Microbiome Forensic Biobanking: A Step toward Microbial Profiling for Forensic Human Identification. *Healthcare* **2021**, *9*, 1371. https://doi.org/10.3390/healthcare9101371

Academic Editor: Francesco Sessa

Received: 15 September 2021
Accepted: 11 October 2021
Published: 14 October 2021

1. Introduction

A biobank is a facility that stores biological material, i.e., tissues and liquids of a living organism—human, animal, plant or microorganism, and collected data, through activities ranging from collection to distribution.

Research biobanks are thus collections of biological material linked with donor data, in particular with clinical and epidemiological data, which can be used for a variety of research projects. The creation of biobanks is closely linked to the development of individualized or personalized medicine [1].

The priority objective of personalized medicine is to produce more effective and better adapted therapies for the patient by linking a huge amount of genomic data with individual health data (diseases, therapies, etc.) and other personal data (e.g., on lifestyle, eating habits, physical activity, income, etc.) [2]. This evolution, made possible by developments in genetic analysis and bioinformatics, has for some years been considered a revolution in medicine [3,4]. Public biobanks belong to, or operate on behalf of, public institutions, notably university hospitals, and are financed by the state. In general, they can be distinguished between tissue banks of the pathology units of universities and central hospitals and national biobanks such as those in the United Kingdom [5], Taiwan [6], or Estonia [7]. Public biobanks are not for profit and operate in the spirit of public service for research, while private biobanks are primarily composed of collections of samples and data collected by pharmaceutical companies and clinical research organizations and essentially come from clinical studies. Small biotech and life sciences companies also set up sample collections for specific research projects. There are private biobanks that pursue commercial purposes and provide their samples and data for a fee to researchers or other companies, which offer direct-to-consumer tests and whose residual samples are withheld, subject to participants' consent, for research purposes [8]. Currently, there are also private–public partnership models in which private investors participate in public biobanks or create and manage biobanks together with public operators [9–11].

For research, it is essential that access to samples and data stored in biobanks is simple and open. Indeed, large quantities of samples and data are required in order to obtain statistically relevant results. It is therefore important that biobanks collaborate across borders and provide data in standardized formats so that they can be compared [12]. Over the years, platforms and networks have emerged that provide researchers with information on the collections and data of the various biobanks, facilitate their access, and eventually link collections together [13].

The benefits of a biobank depend entirely on the research projects for which the samples and data are made available. The overall benefit of the donated samples or of the residual part of samples and the respective data corresponds hypothetically to the scientific value and practical importance of all the research projects for which these samples are used. The overall benefit cannot be determined in concrete terms, but only described in a very abstract way by defining the research areas or typology of research projects on which a biobank is focused. This means that potential donors should be informed about the type of research project supported (purpose of the biobank) and about the project selection process, so that they can evaluate the potential usefulness of a donation [14].

2. The Challenge of Human Microbiome Research

The study of the human microbiome, the genetic heritage of the bacterial communities present within the human body, has always been considered a difficult task [15]. Microbiota are defined as the population of microorganisms (bacteria, fungi, protozoa, and viruses) that colonize an environment in a given time. The microbiome is the totality of the genetic heritage expressed by the microbiota. Every living organism has its own genome and its own genetic heritage; the microbiome is the genome of the microbiota, the genetic heritage of the whole complex of microorganisms present in the organism [16]. The complex structure in which the microbial communities are organized represents an obstacle to traditional in vitro culture, and the sequencing of the microbiome is problematic due to the enormous amount of data to be managed. However, with the development of recent high-throughput sequencing techniques, remarkable progress has been made in the study of the microbiome. It has thus emerged that the microbiome plays a central role, but this is still to be defined in the state of human health, in its metabolism, and in its interaction with drugs.

The microbiome is now considered an essential component of the human biological system [17]. Dynamic, plastic, and variable in the various anatomical sites, during the different stages of life, and in relation to endogenous and exogenous factors, it would seem to play a key role even in numerous clinical conditions [18]. Hence there is a need to deepen our knowledge of a world not yet fully explored insofar as the analysis of microbial communities can represent a useful diagnostic and therapeutic tool [19].

The research conducted in this area is based on the pursuit of different objectives, from the study of the composition and functional properties of the microbiome to its complex dynamics and its interaction with the organism that hosts it [20,21].

Most of the studies published on the microbiome concern the gut microbiome since it is probably the richest one [22] from a biodiversity standpoint.

The creation of microbiome biobanks can promote research with the purpose of adding a fundamental contribution to the understanding of the relationship between microbiota and diseases [23].

Characterizing and studying the human microbiome means analyzing the genetic material of the microbiota and, to this end, there are two common steps to follow: the first step consists of processing the biological samples and extracting the DNA, while it is later sequenced in order to find the order of the nucleic bases along the DNA chain.

In recent years, there have been rapid advances in molecular sequencing and computational techniques. The spread of application of next-generation sequencing (NGS) or high-throughput sequencing (HTS), has incredibly improved the amount of sequencing data that may also be available for forensic purposes. Using NGS to sequence total DNA obtained from a sample allows the sequencing of the whole genome of a given microorganism and the examination of whole communities of microbes, thus obtaining an overview of the resident microbial population.

The main advantages of next-generation sequencing techniques in order to obtain microbiome profiles from samples and individuals are the high levels of parallelism (hundreds of millions of sequential reads in parallel) and the low costs for the production of the DNA sequences [24]. The identification of the taxonomic membership of the components of microbial communities is also of fundamental importance in the study of the microbiome. For this purpose, comparisons are made between the reads and a database that catalogs the association between a certain genome and a particular taxonomic level.

For decades, microbiology has been almost entirely culture-dependent and early studies of the human microbiome involved the culturing of the microbes. Prior to the advent of NGS technologies, only a limited set of information about the human microbiome was available. Nowadays, there are two approaches that are most frequently used in human microbiome research: amplicon sequencing, which relies on sequencing of taxonomic marker genes (usually 16S ribosomal RNA (rRNA) genes) for bacteria and archaea and metagenomic shotgun sequencing, which simultaneously captures all genetic material, providing sequence information on a randomly picked set of DNA fragments extracted from the sample [25]. The 16S rRNA gene, which encodes the small subunit of the bacterial ribosome, is characterized by species-specific variable regions, which are useful for identifying phylogenetic relationships, and highly similar sequences are grouped into operational taxonomic units (OTUs). The assignment of sequences to OTUs is referred to as binning, performed by one of the following methods: unsupervised clustering of similar sequences, phylogenetic models embedding mutation rates and evolutionary relationships, and supervised methods that assign sequences to taxonomic bins based on labeled training data [26]. The bacterial community can be described in terms of which OTUs are represented, their relative abundance, and/or their phylogenetic relationships. The second culture-independent method is metagenomic shotgun sequencing, which simultaneously captures all the genetic material of a sample, providing sequence information on a randomly selected set of DNA fragments, sequencing all microbial genomes within a sample. The information obtained with this method can be used similarly to a 16S rRNA amplicon sequence to identify which taxa are present and the relative abundance of each, and analyze the functional potential of the microbial community. Using this approach, the possibility of detecting species or strain-specific markers is greatly reduced because the larger the genome(s) characterized, the less read depth would be obtained for any particular site [27].

3. The Human Microbiome for Forensic Identification Purposes

While the establishment and development of micobiome research biobanks for clinical applications is already very structured, the development of studies on the applicability of microbiome biobanks for forensic purposes is still in its infancy.

Since the beginning of the 1990s it has been thought that the possibility of typing the genome of bacteria with PCR techniques could be a valid tool to be applied in the forensic field and later, in the 2000s, with the birth of bioterrorism, the applications of forensic microbiology started becoming more established and widespread [28]. An increasing number of scientific contributions have, in the last few years, suggested that in some cases the determination of microbial profiles can be used as forensic evidence or, alternatively, as a complementary element to more traditional forensic evidence [29–31].

Several studies have suggested that an individual can be identified based on profiles of autochthonous microbes that permanently colonize his/her body [32,33]. However, the level of accuracy in identifying a subject decreases as the number of comparison individuals increases since individuals who share the same environment or the same lifestyles could also share the same microbial patterns [34,35].

Since human identification is a comparative analysis, the microbial trace should be compared with a reference sample constituted of site-related microbial communities in order to be linked to the person who left it behind. In the past few years, several studies have focused on stability in the human microbiome over time [36]. The literature published in this field over the past few years has ascertained clearly enough that microbial communities, although personalized, vary systematically across body sites and time, with intrapersonal differences over time being smaller than interpersonal ones, showing such a high degree of spatial and temporal variability that the degree and nature of this variability can constitute in and of itself an important parameter that is useful in distinguishing individuals from one another [37–41]. It is paramount to make the effort to organically synthesize all results achieved until now and to implement the number of participants to this type of study. Therefore, the observations of the available literature warrant further studies to enable forensic scientists to evaluate the risk of microbial transfer and recovery from various items and to further critically evaluate the suitability of current human DNA recovery protocols for microbial profiling, or establish new protocols to prevent microbial transfer and contamination of forensic evidence [42,43].

Therefore, in order to increase knowledge in this fascinating, and relatively new, field of application of forensic microbiology, the creation of large population microbiome biobanks, specifically dedicated to forensic human identification, could be worthwhile. This could also be useful in increasing the practical applications of forensic microbiology for identification purposes, given that this type of evidence is currently absent from most real casework investigations and judicial proceedings in courts.

4. Identity and Forensic Identification

Although personhood and identity have never been simple concepts, as we learn more about ourselves as an amalgam of us and the microbes that live on us and within us, we will rethink our concepts of personal identity and normalcy. Our understanding of the human microbiome and its interaction with the human body also has implications for how we conceptualize personal identity. If we consider that each individual's microbiome is unique, then the microbiome may be incorporated into how we define ourselves as people, and thus it would be important to clarify whether it is possible to identify a person through his/her microbiome.

The concepts of identification and therefore identity are crucial in forensics, since forensic sciences are commonly aimed at identifying people, toxic substances, and objects. It is abundantly clear that each individual is characterized by physical and biological particularities that distinguish him/her from any other individual and even make him/her unique and unrepeatable within the human species. These differentiations are all the more important the greater the demographic development of the human community of reference, much more so when the community itself becomes a civil society and therefore subject to a legal order. In this case, physical identification corresponds to legal identification, virtually reconstructed, according to two-way reference models relating to the anthropomorphic characteristics and the assignment of personal details, as recognized by the particular legal system that has jurisdiction over the person in question. In the case of Western societies, for example, they would include the name, date of birth, place of birth, citizenship, sex, and some other traits that characterize the individual. If such legal characterizations did not exist, it is easy to understand how any legal relationship and, from a sociological point of view, any type of relationship between individuals would fail, lacking the certainty of the interlocutor's identifiability. The term identity derives from the medieval term identicus, which in turn is of Latin derivation, with the meaning of "the same".

When referring to personal identity, we talk about how to understand and explain how a person can remain the same despite the physical, psychological, existential changes he/she goes through throughout his/her life. The concept of identity in a more scientific sense is made clearer by resorting to the principle of absolute identity, according to which everything is identical to itself and can only be itself. This means that identity is the quality of a thing that makes it the only one and differentiates itself from everything else. At this point, it is normal to ask how it is possible to make a comparison between two individualities if, by definition, the principle of absolute identity prevents us from doing so: the answer is given to us by the principle of relative identity, i.e., the possibility of comparison existing between two terms of comparison which, although being an expression of two distinct individualities (in an absolute sense), can also be considered identical to each other, as ways of being of the same reality.

It is clear that the identity we are now talking about in the forensic field is not to be understood as absolute, but rather relative. In fact, what we are dealing with is nothing other than the comparison between two distinct absolute individualities (think of two genetic profiles to be compared, one of which belongs to a certain person) that are expressions of the same reality [44]. This can be considered the core difference between classification (placing an object in a defined category) and identification (the recognition of uniqueness—that something is one of a kind). Therefore, the term identification refers to the technical-scientific activity aimed at establishing the identity of any material and, in the case of personal identification, the identity of a person. Identity in itself encompasses two concepts: the comparison between two terms and the resulting judgment. No individual can be identified with another; that is, being him/herself and at the same time another subject; it is only itself and is identical to itself only in the instant in which it is observed. An individualization can be viewed as a special case of identification, where the restricted class is populated by only one object. Definitions of individualization in the forensic literature (e.g., fingerprint, footwear marks, or tool marks) systematically refer to the capability of pointing to the right source to the exclusion of all others (objects or persons) [45]. Hence, by default, the size of the population of relevant sources considered at the outset of the examination is systematically set to its maximum, regardless of the specific circumstances of the case. We call this the Earth population paradigm. In that paradigm, the individualization conclusion cannot be reached in a deductive manner, but is de facto probabilistic in nature.

5. Human Microbiome Biobanking for Forensic Research in Human Identification

Today, in forensic genetics, human identification is always performed following specific investigative directives which involve observation of the proof, collection, analysis, and scientific interpretation of the results. Once identified, the biological material is collected, stored, and eventually characterized following specific procedures. Then, in the laboratory, DNA is extracted and quantified. Current forensic methods typically rely on targeting genetic markers to create genetic profiles to compare evidentiary items with profiles generated from a reference sample from an individual. With this purpose, specific DNA regions are targeted and amplified by PCR reaction using commercial kits that allow multiplex amplification of specific sets (multiplex) of short tandem repeat (STR) markers, insertion/deletion polymorphisms, or other markers useful for forensic aims. In some cases when the evidentiary sample may be degraded or contain low amounts of DNA (i.e., low-copy number (LCN) DNA), high-copy number (HCN) markers (i.e., the mitochondrial genome or hypervariable regions of the mitochondrial genome) are targeted.

Other HCN markers, such as skin microbiome genetic markers, may provide additional identifying genetic information that can be used independently or potentially in conjunction with partial human forensic marker profiles [39].

Given that human identification is comparative in its nature, when using the microbiome for identification purposes, the microbial profile has to be compared with that of a reference sample of interest in order to be linked to a specific person. In using the microbiome for human identification, it should be considered that the microbiome varies among different tissues and body sites and also over time, although the variability between individuals seems to be greater than the variability between different parts of the same person's body [42]. This degree of variability constitutes in itself an important element in being able to distinguish the subjects from each other and, therefore, in considering the determination of the microbiome as a useful technique, in association with other more consolidated techniques, for personal identification in the forensic field [46].

The application of these new technologies consists of gaining information about the microbiome profile of a wanted person from a trace itself. Its use could be especially useful in investigative cases where there are no potential suspects and no match between the evidence DNA sample under investigation and any genetic profiles entered, for example, in criminal databases [47,48]. Through microbiome prediction starting from biological samples found at the crime scene, probabilistic information may be acquired as to the same characteristics of the sample donor, such as past exposures, visits to other countries, predisposition to certain conditions, sexual practices, diet, and consumption of tobacco, alcohol, and other drugs [49]. The combination of these elements, therefore, narrows the circle of possible perpetrators and facilitates investigations, adding qualitative information that should be integrated with other investigative elements [50].

Having tens of thousands of samples stored in dedicated forensic biobanks to be used for forensic studies would allow us to achieve very important objectives. In particular, the first aspect to be addressed is the quantification of diversity of intra-subject species, in order to characterize the microbiome and evaluate its variability in different populations, within the subjects, regardless of their state of health. Secondly, it would be essential to establish the inter-subject variability of the microbiome, quantifying the differences between the bacterial communities of different subjects, considering various environmental factors. Finally, the differences between the microbiome of different subjects, with different lifestyles, could be quantified, in order to evaluate the reliability of the prediction on the lifestyles of those who left a trace starting from the analysis of the microbiome [51].

The creation of this type of "forensic microbiome biobank" would undoubtedly allow for the improvement of technologies for the isolation and analysis of bacterial organisms, the development of a set of sequences of reference microbial genome and the development of new tools for computational analysis and new technologies for sequencing (to be associated with those already in use in many forensic laboratories), which permit the examining of the genome of bacterial communities since the creation of numerous and complex data sets always requires new analysis tools. Above all, given the fact that no standardized technique is available for the purpose of microbiome forensic profiling, it is important to obtain comparable data from different studies. Therefore, the definition of shared objectives, strategies, and protocols in this area represents a precious and irreplaceable opportunity to enrich, compare, and consolidate skills, creating a constantly updated heritage for researchers and forensic scientists [52]. To reach this goal, it is necessary that the institution of biobanks of samples and data also contains reference databases categorized by, for example, type of sample, individual age, race, habits, and ethnicity in order to detect the possibility of determining a "core microbiome" among human subjects. Furthermore, the institutionalization of this type of research within dedicated biobanks may allow for the establishment of one or more centers for the coordination and analysis of data, which manages the processed and unprocessed data, coordinates the analyses and establishes a portal through which it is possible to give international visibility to projects and support international relations. Moreover, thanks to these structures, resources can be widely available and accessible to the scientific community.

It is also important to have facilities that allow biological material to be stored in the best way to ensure the reproducibility and comparability of research results—above all, because the goal should be to have analytical standards, reference databases, and predictive methods to be applied in concrete cases of judicial investigation. The essential standards are represented by the horizontal requirements for the infrastructure, the competence of the staff of a biobank, the quality management system (QMS), the equipment, the quality control (QC), and the procedures for the management, sample processing, and storage, including method validation and verification.

In the case of microbiome forensic biobanks, samples will be obtained through non-invasive or minimally invasive means, for example, including skin/brushes, oral swabs, saliva samples, nasal swabs, vaginal swabs, and self-collected fecal samples. Moreover, leftover materials collected during endoscopies to collect the gut microbiome may add a minimal additional risk to conducting this type of research [53]. Because the risks of most human microbiome research and biobanks are often negligible, they involve only the lowest measure of "minimal risk" as defined in many regulations. Rhodes et al. propose a new conception and category of risk, that is "de minimis risk", to appropriately describe the risks in the context of human microbiome research. As they explained, "it entails a degree of risk so low that harms are nominal and unlikely" [54]. However, as we gain a greater understanding of variation in the microbiota that inhabit different parts of the body, as well as the advantages of deep versus minimally invasive sampling, sampling techniques and associated risk–benefit assessments may change.

The establishment of this type of biobank, due to the very delicate nature of the activities connected to it, should also meet certain regulatory requirements and quality standards that guarantee its correct functioning, impartiality, the presence of all the requirements, and the protection of donors' personal data. The ethical issues and logistical challenges arising with the use of microbiome biobanks vary with the nature of the research. From a general point of view, we can assume that these are similar to the many concerns raised by other types of biobanks.

The social, ethical, and political concerns and issues pertaining to microbiome identification in forensics are situated within the intersection of civil rights, science, and governance. They are intimately linked to the constitution of new and wider groups of populations as "microbiome suspects". Such concerns include, but are not limited to, privacy, surveillance, ideological and scientific interpretation of such evidence, and the scientific reliability of microbiome identification, as well as the potential misuse and abuse of criminal investigations. However, the fact that human microbiome research samples also contain human DNA raises concerns about privacy and confidentiality, since these samples can be analyzed in ways that are identifiable. In this respect, we suggest that human microbiome research samples should be treated by biobanks with the same safeguards in terms of privacy and confidentiality as any other human tissue samples or identifying sources of information.

6. Conclusions

The possibility of using the human microbiome as a means of personal identification undoubtedly represents a fascinating and innovative field of research. The establishment of dedicated biobanks for the development of research in this area would allow for the obtaining of reproducible and robust results to allow these techniques to be used in the very near future, even in real criminal cases. Applying new technologies or considering new applications of technologies that have already been consolidated in the forensic and judicial field means proceeding on two tracks: on the one hand, research, to update knowledge; and on the other, to keep in consideration the regulatory and ethical aspects.

Author Contributions: Conceptualization, P.T. and L.C.; methodology, P.T.; investigation, P.T.; resources, P.T. and L.C.; writing—original draft preparation, P.T.; writing—review and editing, L.C.; supervision, L.C. Both authors have read and agreed to the published version of the manuscript.

Funding: This research received no external funding.

Institutional Review Board Statement: Not applicable.

Informed Consent Statement: Not applicable.

Data Availability Statement: No new data were created or analyzed in this study. Data sharing is not applicable to this article.

Conflicts of Interest: The authors declare no conflict of interest.

References

1. Mandl, K.D.; Glauser, T.; Krantz, I.D.; Avillach, P.; Bartels, A.; Beggs, A.H.; Biswas, S.; Bourgeois, F.T.; Corsmo, J.; Dauber, A.; et al. Genomics Research and Innovation Network. The Genomics Research and Innovation Network: Creating an interoperable, federated, genomics learning system. *Genet. Med.* **2020**, *22*, 371–380. [CrossRef]
2. Lewis, C.; McQuaid, S.; Hamilton, P.W.; Salto-Tellez, M.; McArt, D.; James, J.A. Building a 'Repository of Science': The importance of integrating biobanks within molecular pathology programmes. *Eur. J. Cancer* **2016**, *67*, 191–199. [CrossRef]
3. Hartman, V.; Matzke, L.; Watson, P.H. Biospecimen Complexity and the Evolution of Biobanks. *Biopreserv. Biobank.* **2019**, *17*, 264–270. [CrossRef] [PubMed]
4. Coppola, L.; Cianflone, A.; Grimaldi, A.M.; Incoronato, M.; Bevilacqua, P.; Messina, F.; Baselice, S.; Soricelli, A.; Mirabelli, P.; Salvatore, M. Biobanking in health care: Evolution and future directions. *J. Transl. Med.* **2019**, *17*, 172. [CrossRef] [PubMed]
5. Sudlow, C.; Gallacher, J.; Allen, N.; Beral, V.; Burton, P.; Danesh, J.; Downey, P.; Elliott, P.; Green, J.; Landray, M.; et al. UK biobank: An open access resource for identifying the causes of a wide range of complex diseases of middle and old age. *PLoS Med.* **2015**, *12*, e1001779. [CrossRef] [PubMed]
6. Lin, J.C.; Hsiao, W.W.; Fan, C.T. Transformation of the Taiwan Biobank 3.0: Vertical and horizontal integration. *J. Transl. Med.* **2020**, *18*, 304. [CrossRef] [PubMed]
7. Leitsalu, L.; Alavere, H.; Tammesoo, M.L.; Leego, E.; Metspalu, A. Linking a population biobank with national health registries-the estonian experience. *J. Pers. Med.* **2015**, *5*, 96–106. [CrossRef]
8. Tozzo, P.; Caenazzo, L. The Skeleton in the Closet: Faults and Strengths of Public versus Private Genetic Biobanks. *Biomolecules* **2020**, *10*, 1273. [CrossRef]
9. Karimi-Busheri, F.; Rasouli-Nia, A. Integration, Networking, and Global Biobanking in the Age of New Biology. *Adv. Exp. Med. Biol.* **2015**, *864*, 1–9. [CrossRef]
10. Bledsoe, M.J.; Sexton, K.C. Ensuring Effective Utilization of Biospecimens: Design, Marketing, and Other Important Approaches. *Biopreserv. Biobank.* **2019**, *17*, 248–257. [CrossRef]
11. Caenazzo, L.; Tozzo, P. The Future of Biobanking: What Is Next? *BioTech* **2020**, *9*, 23. [CrossRef]
12. Caenazzo, L.; Tozzo, P.; Borovecki, A. Ethical governance in biobanks linked to electronic health records. *Eur. Rev. Med. Pharmacol. Sci.* **2015**, *19*, 4182–4186.
13. Henderson, M.K.; Matharoo-Ball, B.; Schacter, B.; Kozlakidis, Z.; Smits, E.; Törnwall, O.; Litton, J.E. Global Biobank Week: Toward Harmonization in Biobanking. *Biopreserv. Biobank.* **2017**, *15*, 491–493. [CrossRef] [PubMed]
14. Gefenas, E.; Dranseika, V.; Serepkaite, J.; Cekanauskaite, A.; Caenazzo, L.; Gordijn, B.; Pegoraro, R.; Yuko, E. Turning residual human biological materials into research collections: Playing with consent. *J. Med. Ethics* **2012**, *38*, 351–355. [CrossRef] [PubMed]
15. Human Microbiome Project Consortium. A framework for human microbiome research. *Nature* **2012**, *486*, 215–221. [CrossRef]
16. Marchesi, J.R.; Ravel, J. The vocabulary of microbiome research: A proposal. *Microbiome* **2015**, *3*, 31. [CrossRef]
17. Dominguez-Bello, M.G.; Godoy-Vitorino, F.; Knight, R.; Blaser, M.J. Role of the microbiome in human development. *Gut* **2019**, *68*, 1108–1114. [CrossRef] [PubMed]
18. Young, V.B. The role of the microbiome in human health and disease: An introduction for clinicians. *BMJ (Clin. Res. Ed.)* **2017**, *356*, j831. [CrossRef]
19. Cong, J.; Zhang, X. How human microbiome talks to health and disease. *Eur. J. Clin. Microbiol. Infect. Dis.* **2018**, *37*, 1595–1601. [CrossRef]
20. Aagaard, K.; Petrosino, J.; Keitel, W.; Watson, M.; Katancik, J.; Garcia, N.; Patel, S.; Cutting, M.; Madden, T.; Hamilton, H.; et al. The Human Microbiome Project strategy for comprehensive sampling of the human microbiome and why it matters. *FASEB J.* **2013**, *27*, 1012–1022. [CrossRef]
21. Blum, H.E. The human microbiome. *Adv. Med. Sci.* **2017**, *62*, 414–420. [CrossRef]
22. Fan, Y.; Pedersen, O. Gut microbiota in human metabolic health and disease. *Nat. Rev. Microbiol.* **2021**, *19*, 55–71. [CrossRef]
23. Barzegari, A.; Saeedi, N.; Saei, A.A. Shrinkage of the human core microbiome and a proposal for launching microbiome biobanks. *Future Microbiol.* **2014**, *9*, 639–656. [CrossRef]
24. Malla, M.A.; Dubey, A.; Kumar, A.; Yadav, S.; Hashem, A.; Abd Allah, E.F. Exploring the Human Microbiome: The Potential Future Role of Next-Generation Sequencing in Disease Diagnosis and Treatment. *Front. Immunol.* **2019**, *9*, 2868. [CrossRef]
25. Knight, R.; Vrbanac, A.; Taylor, B.C.; Aksenov, A.; Callewaert, C.; Debelius, J.; Gonzalez, A.; Kosciolek, T.; McCall, L.I.; McDonald, D.; et al. Best practices for analysing microbiomes. *Nat. Rev. Microbiol.* **2018**, *16*, 410–422. [CrossRef]
26. Galloway-Peña, J.; Hanson, B. Tools for Analysis of the Microbiome. *Dig. Dis. Sci.* **2020**, *65*, 674–685. [CrossRef] [PubMed]
27. Franzosa, E.A.; Huang, K.; Meadow, J.F.; Gevers, D.; Lemon, K.P.; Bohannan, B.J.; Huttenhower, C. Identifying personal microbiomes using metagenomic codes. *Proc. Natl. Acad. Sci. USA* **2015**, *112*, E2930–E2938. [CrossRef] [PubMed]
28. Budowle, B.; Schutzer, S.E.; Einseln, A.; Kelley, L.C.; Walsh, A.C.; Smith, J.A.L.; Marrone, B.L.; Robertson, J.; Campos, J. Public health. Building microbial forensics as a response to bioterrorism. *Science* **2003**, *301*, 1852–1853. [CrossRef] [PubMed]
29. Meadow, J.F.; Altrichter, A.E.; Bateman, A.C.; Stenson, J.; Brown, G.Z.; Green, J.L.; Bohannan, B.J. Humans differ in their personal microbial cloud. *PeerJ* **2015**, *3*, e1258. [CrossRef]

30. Quaak, F.; van Duijn, T.; Hoogenboom, J.; Kloosterman, A.D.; Kuiper, I. Human-associated microbial populations as evidence in forensic casework. *Forensic Sci. Int. Genet.* **2018**, *36*, 176–185. [CrossRef]
31. Singh, H.; Clarke, T.; Brinkac, L.; Greco, C.; Nelson, K.E. Forensic Microbiome Database: A Tool for Forensic Geolocation Meta-Analysis Using Publicly Available 16S rRNA Microbiome Sequencing. *Front. Microbiol.* **2021**, *12*, 644861. [CrossRef] [PubMed]
32. Clarke, T.H.; Gomez, A.; Singh, H.; Nelson, K.E.; Brinkac, L.M. Integrating the microbiome as a resource in the forensics toolkit. *Forensic Sci. Int. Genet.* **2017**, *30*, 141–147. [CrossRef] [PubMed]
33. Schmedes, S.E.; Woerner, A.E.; Budowle, B. Forensic Human Identification Using Skin Microbiomes. *Appl. Environ. Microbiol.* **2017**, *83*, e01672-17. [CrossRef]
34. Schmedes, S.E.; Woerner, A.E.; Novroski, N.; Wendt, F.R.; King, J.L.; Stephens, K.M.; Budowle, B. Targeted sequencing of clade-specific markers from skin microbiomes for forensic human identification. *Forensic Sci. Int. Genet.* **2018**, *32*, 50–61. [CrossRef]
35. Phan, K.; Barash, M.; Spindler, X.; Gunn, P.; Roux, C. Retrieving forensic information about the donor through bacterial profiling. *Int. J. Leg. Med.* **2020**, *134*, 21–29. [CrossRef]
36. Costello, E.K.; Lauber, C.L.; Hamady, M.; Fierer, N.; Gordon, J.I.; Knight, R. Bacterial community variation in human body habitats across space and time. *Science* **2009**, *326*, 1694–1697. [CrossRef] [PubMed]
37. D'Angiolella, G.; Tozzo, P.; Gino, S.; Caenazzo, L. Trick or Treating in Forensics—The Challenge of the Saliva Microbiome: A Narrative Review. *Microorganisms* **2020**, *8*, 1501. [CrossRef]
38. Tozzo, P.; D'angiolella, G.; Brun, P.; Castagliuolo, I.; Gino, S.; Caenazzo, L. Skin microbiome analysis for forensic human identification: What do we know so far? *Microorganisms* **2020**, *8*, 873. [CrossRef]
39. Procopio, N.; Lovisolo, F.; Sguazzi, G.; Ghignone, S.; Voyron, S.; Migliario, M.; Renò, F.; Sellitto, F.; D'Angiolella, G.; Tozzo, P.; et al. "Touch microbiome" as a potential tool for forensic investigation: A pilot study. *J. Forensic Leg. Med.* **2021**, *82*, 102223. [CrossRef]
40. García, M.G.; Pérez-Cárceles, M.D.; Osuna, E.; Legaz, I. Impact of the Human Microbiome in Forensic Sciences: A Systematic Review. *Appl. Environ. Microbiol.* **2020**, *86*, e01451-20, PMID: 32887714; PMCID: PMC7642070. [CrossRef]
41. Neckovic, A.; van Oorschot, R.; Szkuta, B.; Durdle, A. Investigation of direct and indirect transfer of microbiomes between individuals. *Forensic Sci. Int. Genet.* **2020**, *45*, 102212. [CrossRef] [PubMed]
42. Neckovic, A.; A H van Oorschot, R.; Szkuta, B.; Durdle, A. Challenges in Human Skin Microbial Profiling for Forensic Science: A Review. *Genes* **2020**, *11*, 1015. [CrossRef]
43. Dumache, R.; Ciocan, V.; Muresan, C.; Enache, A. Molecular DNA Analysis in Forensic Identification. *Clin. Lab.* **2016**, *62*, 245–248. [CrossRef]
44. Arenas, M.; Pereira, F.; Oliveira, M.; Pinto, N.; Lopes, A.M.; Gomes, V.; Carracedo, A.; Amorim, A. Forensic genetics and genomics: Much more than just a human affair. *PLoS Genet.* **2017**, *13*, e1006960. [CrossRef]
45. Hampton-Marcell, J.T.; Larsen, P.; Anton, T.; Cralle, L.; Sangwan, N.; Lax, S.; Gottel, N.; Salas-Garcia, M.; Young, C.; Duncan, G.; et al. Detecting personal microbiota signatures at artificial crime scenes. *Forensic Sci. Int.* **2020**, *313*, 110351. [CrossRef]
46. Schmedes, S.E.; Sajantila, A.; Budowle, B. Expansion of Microbial Forensics. *J. Clin. Microbiol.* **2016**, *54*, 1964–1974. [CrossRef]
47. Kuiper, I. Microbial forensics: Next-generation sequencing as catalyst: The use of new sequencing technologies to analyze whole microbial communities could become a powerful tool for forensic and criminal investigations. *EMBO Rep.* **2016**, *17*, 1085–1087. [CrossRef]
48. Metcalf, J.L.; Xu, Z.Z.; Bouslimani, A.; Dorrestein, P.; Carter, D.O.; Knight, R. Microbiome Tools for Forensic Science. *Trends Biotechnol.* **2017**, *35*, 814–823. [CrossRef] [PubMed]
49. Robinson, J.M.; Pasternak, Z.; Mason, C.E.; Elhaik, E. Forensic Applications of Microbiomics: A Review. *Front. Microbiol.* **2021**, *11*, 608101. [CrossRef]
50. Oliveira, M.; Amorim, A. Microbial forensics: New breakthroughs and future prospects. *Appl. Microbiol. Biotechnol.* **2018**, *102*, 10377–10391. [CrossRef] [PubMed]
51. Pechal, J.L.; Schmidt, C.J.; Jordan, H.R.; Benbow, M.E. A large-scale survey of the postmortem human microbiome, and its potential to provide insight into the living health condition. *Sci. Rep.* **2018**, *8*, 5724. [CrossRef] [PubMed]
52. Tozzo, P.; Angiola, F.; Caenazzo, L. Letter to the Editor. Progress in morgues: The time has come for a wide network of forensic research biobanks. *Int. J. Leg. Med.* **2021**, *135*, 2135–2137. [CrossRef] [PubMed]
53. Caenazzo, L.; Tozzo, P.; Pegoraro, R. Biobanking research on oncological residual material: A framework between the rights of the individual and the interest of society. *BMC Med. Ethics* **2013**, *14*, 17. [CrossRef]
54. Rhodes, R.; Azzouni, J.; Baumrin, S.B.; Benkov, K.; Blaser, M.J.; Brenner, B.; Dauben, J.W.; Earle, W.J.; Frank, L.; Gligorov, N.; et al. De minimis risk: A proposal for a new category of research risk. *Am. J. Bioeth.* **2011**, *11*, 1–7. [CrossRef]

 *healthcare* 

Case Report

Two Cases of Feigned Homicidality: Assessing the Third Dimension in Homicidal Threats

Pavan Madan [1,*], Alexander Graypel [2] and Alan R. Felthous [3]

[1] Mindpath Health, Davis, CA 95616, USA
[2] Community and Long-Term Care Psychiatry LLC, Saint Louis, MO 63128, USA
[3] Department of Neurology and Psychiatry, Saint Louis University, Saint Louis, MO 63103, USA
* Correspondence: Pmadan@cpsych.com

Abstract: Although data and research on the topic are lacking, the phenomenon of feigned homicidality in short-term hospitalization appears to have increased in recent years. Inpatient psychiatrists not only assess the seriousness of homicidal threats, but also whether such threats are authentic. However, specific literature and diagnostic manuals provide virtually no clinical guidance for this. The authors present two case examples of homicidality feigned for self-serving purposes that had little to do with hostility against the would-be victim. They recommend an approach to assessment that first takes any threat of homicide seriously, and involves an attempt to assess the seriousness of the threat and risk of harm. Secondly, if feigned homicidality is suspected, clinicians can methodically assess for this using criterion that have been applied to the assessment of malingering.

Keywords: homicidal; case series; malingering; feigned; threat; risk assessment; psychiatric evaluation

Citation: Madan, P.; Graypel, A.; Felthous, A.R. Two Cases of Feigned Homicidality: Assessing the Third Dimension in Homicidal Threats. *Healthcare* **2022**, *10*, 31. https://doi.org/10.3390/healthcare10010031

Academic Editors: Francesco Sessa and Mariano Cingolani

Received: 10 November 2021
Accepted: 21 December 2021
Published: 24 December 2021

Publisher's Note: MDPI stays neutral with regard to jurisdictional claims in published maps and institutional affiliations.

1. Introduction

The professional literature on assessment of a patient's verbal threat to kill concerns a determination as to whether the homicidal statement is only a fantasy, not likely to be carried out, or a serious intention with a realistic potential for a lethal or injurious outcome. Not usually included in such discussions is a third possibility: the verbal statement is made to achieve a self-serving but non-homicidal goal. In many cases, the possibility of feigned homicidality must be considered together with fantasized and planned homicide when assessing risk. Without research and scientific literature on feigned homicidality, the clinician is faced with a daunting task when a patient threatens to kill another person.

Sixty years ago, it would have been rare for anyone to threaten to kill another person simply to gain hospital access. Times were different. Involuntary hospitalization in a mental facility could extend for months or years. Homelessness of the mentally ill was limited, and addiction to street drugs was not as prevalent. In recent years, increasing reference is made to the patient who achieves hospitalization for "three hots and a cot", rather than for psychiatric treatment of symptoms of mental illness [1]. Without mentioning feigned homicidality specifically, literature on malingering recognizes hospitalization as a goal in some cases. However, feigned homicidality seems to be becoming a more frequent strategy for gaining hospitalization for various reasons.

Patients often have alternative motives for entering hospitals, such as the motive to detoxify comfortably from addicting drugs, to seek sedative medications, to obtain lodging that is more comfortable than a homeless shelter, or to avoid jail detainment. Intended escape from destructive conditions and circumstances is a positive motivation that is adaptive and supportive of mental health. Once detoxified, the addicted patient no longer claims suicidal or homicidal thoughts and may even admit to having fabricated such thoughts in order to gain hospitalization. Within a few days, they are eager to return to the situation from which they came without considering rehabilitative measures.

In this article, we present two cases of feigned homicidal ideations to gain access to hospital. We then discuss the recommended evaluation of verbal threat, assessment of risk, differentiation from pseudologia fantastica, and approach towards cases with suspected feigning of homicidal ideation. We hope that the cases and the discussion provide the reader with an understanding of how such cases may present at different stages of mental health treatment in a hospital, and when feigned homicidality is suspected, how the providers can perform a reasonable assessment to establish a diagnosis and treatment plan. We use two unrelated cases as examples. Both cases are discussed at length below, and their main findings are later summarized in Table 1.

Table 1. Summary of pertinent findings in the two cases that point towards a possibility of feigned homicidal ideations.

CASE 1	CASE 2
History of substance abuse and legal problems	History of substance abuse
Vague symptoms	Symptoms changed based on circumstances
Uncooperative with detailed examination	Lack of clear plan, means, or history of violence
Story changing over time	Inconsistencies in history
Medical examination did not match history	Irritable when confronted with inconsistencies
Collateral information from family contradicted patient's story	Collateral information from significant other confirmed that they were feigning psychiatric symptoms
Clear motive to stay inside the hospital and avoid legal consequences during the time of stay	Clear motive to feign symptoms to avoid homelessness and gain a place to stay

2. Case Examples

2.1. CASE I

Mr. A was a Caucasian man in his forties with a history of chronic medical conditions and multiple emergency room (ER) visits for chest pain and alcohol intoxication. He was homeless and unemployed, having recently lost his job as a cook. He had no documented past psychiatric history. He was brought to the ER by emergency medical services after complaining of chest pain and possible overdose. Mr. A reported to the ER physician that he was having suicidal ideation and planned to drink excess amounts of alcohol, hang himself, or shoot himself with a gun, which he said he did not possess. The patient presented symptoms of depressed mood, hopelessness, helplessness, and decreased interest in things as well as decreased sleep. He reported a pending DUI charge, regular alcohol use with blackouts, buildup of tolerance, and withdrawal tremors. He denied symptoms consistent with mania. He denied hearing voices or having visions and showed no signs of response to internal stimuli.

The patient had hypertension, bilateral upper extremity neuropathy, COPD, and angina. He also reported a history of a stroke 6 years earlier, which led to loss of function in his right arm and left leg. Per neurology consultation, this stroke was very unlikely, and the patient had no current stroke-related deficits.

On the mental status examination, he was hostile and irritable during the interview. He described his mood as "very sad at this point". Thought content included suicidal ideation with a method. He denied experiencing voices or visions. He was alert and oriented to place and time and appeared to have medical decision-making capacity, but he refused to cooperate with the mini-mental status exam.

On the first day after admission, the patient said he slept poorly; however, the nurses reported that he had slept for 8 h. He showed no withdrawal symptoms. He continued to express suicidal ideation without any plan. He denied having homicidal thoughts towards anyone. On the second day, he was angry over the food, the nurses, other patients, the medical students, and at the unit in general. He did not attend any of the group therapy sessions, was aggressive during interviews, and reported his mood to be "terrible". He also started complaining of voices telling him to hurt himself; however, he was never seen responding to internal stimuli.

Over the next few days, the patient continued to report suicidal ideation with various methods, ranging from hanging to not having a method. He became more social and attended some groups. During one interview, he revealed that he had homicidal ideation towards five individuals outside the hospital, but he declined to provide any information about them. He displayed narcissistic traits, especially when criticizing the hospital food, often saying "I can cook around these fools". He described his desire to work in a fancy restaurant one day, which was inconsistent with his talk of abruptly bringing about his own death.

When asked for consent to talk to his family, he said that he would allow his doctors to talk to his brother, but only after he talked with him first. Immediately after sharing that his brother was an important government employee, he described elaborate revenge fantasies towards three individuals, who he said had wronged him in the past. One of the individuals was a previous employer, the other individual was his ex- girlfriend's mother, who he said he wanted to torture and strangle. This was to retaliate for causing him and his girlfriend to break up. The third individual used to be his best friend, but then this man impregnated his ex-girlfriend. He claimed that he thought this because he knew his friend's blood type and that of his own, and believed that the baby's blood group corresponded with that of his best friend and his ex-girlfriend's, and therefore, that the baby must have been fathered by the other man. There were two more people who the patient said he had thoughts of killing, but he refused to tell his treatment team about them. He exclaimed dramatically, "I will tell you when the time is right!" Over the next few days, Mr. A attended all of the groups, ate all of his meals, and was cooperative with medication.

Mr. A spoke more about his "revenge plans", which were very unrealistic and sounded like a revenge fantasy movie. He said that he did not expect to get caught because he was "just too good". One of the authors spoke with Mr. A's brother, who confirmed the patient's history of alcoholism. His brother was unsure about any "revenge plans" and could not confirm any of the events that the patient described. However, he shared that his brother had been emotionally hurt by his ex-girlfriend in the past, but he had not heard about any paternity situation. His brother said that the patient had a good support system on the outside and many family members; however, many of them have children, and they did not want Mr. A around them when he was drunk. Once Mr. A becomes abstinent, he would have their full support.

Mr. A then developed plans to enter an alcohol rehabilitation program close to home so he could spend more time with his family. When asked again about homicidal thoughts, he described them but was inconsistent and less detailed than before. He again talked about his ex-girlfriend, but this time, he appeared to have forgotten what he had reported to his treatment team a few days earlier and said that he was very depressed because his ex-girlfriend had died in a car accident with the baby still inside of her. There were several inconsistencies in his story, and his brother also confirmed that the ex-girlfriend was still alive despite the patient having said that she died in a car accident. Other homicidal thoughts continued to be vague, and he continued to alter the number of people he wanted to kill. Citalopram and trazodone were discontinued and replaced with mirtazapine to help with his chronically poor sleep, appetite, and mood. The patient placed a telephone call and planned to go into rehabilitation.

A few days later, the patient asked the physician if he could have a doctor's note confirming his stay in the hospital with the exact dates of hospitalization. He said he needed the note because he had missed two court cases that were scheduled two days following admission, and that he was unable to attend them as a result of being in the hospital. He added that this had nothing to do with the reason that he presented to the hospital.

The next day, the patient said his homicidal thoughts were only towards two people. He was still irritable on the unit and complained about the food and other patients. However, he was social with his peers, laughing, playing cards, and joking at times.

As the time neared the patient's rehabilitation intake date, he began to deny suicidal ideations. He continued to express homicidal ideation, but these thoughts were assessed as likely revenge fantasies or malingering given their inconsistencies and vagueness.

The patient was contacted by one of the authors 3 months after his discharge to see how he was doing. He said that he was doing great and maintaining his sobriety. When asked about his homicidal thoughts, the patient said, "That's the furthest thing from my mind".

The diagnosis of feigned homicidality was based on (1) inconsistent history over time; (2) inconsistency between Mr. A's account from that of his brother; (3) evidence of a clear external incentive for hospitalization; (4) no clear plan for homicide or access to a weapon, and; (5) lack of cooperation with the diagnostic assessment.

2.2. *CASE 2*

Mr. B was a 46-year-old African American divorced male, with a history of schizoaffective disorder and hypertension, who presented to the ER with complaints of "hearing voices and feeling suicidal and homicidal". The patient told the ER physicians that he was hearing voices telling him to jump off a bridge. He said that, one week before, on being commanded by voices, he had jumped off the balcony from his second floor and suffered minor knee abrasions. When the psychiatry resident interviewed Mr. B, he said that the voices were telling him to jump off the roof of his two-story apartment (but he did not mention the bridge). He said that he was hearing male voices telling him to hurt his sister and kill her. He said that he was upset with his sister due to a will dispute, which arose after the death of their mother 2 years previously. He alleged that his sister took money from the inheritance to open a business, and that this incident had infuriated all of his siblings. Mr. B stated that he did not see her on a regular basis, but knew where she lived and how to reach her.

Mr. B reported several depressive symptoms including feeling depressed, poor sleep, poor appetite, 15 lbs. weight loss, feelings of loneliness, hopelessness, anger at God for taking away his mother, poor concentration, and lack of interest in pleasurable activities. He also reported feeling nervous. He reported feeling paranoid that "people are out there to get me".

Mr. B reported that he had been previously diagnosed with schizophrenia, bipolar disorder and schizoaffective disorder. He stated that he first started hearing voices at the age of 32 years, after the death of his father. He had been taking antipsychotics off and on for the past several years. Previous psychiatric hospitalization revealed the diagnoses of cannabis abuse, cocaine abuse, and substance-induced mood and psychotic disorders. The patient reported having had multiple psychiatric hospitalizations in the past, including one in the same hospital as the current presentation. He reported to have made one suicide attempt 4 years earlier by overdosing on his medications "because the voices told me to do so" (he did not report any suicide attempt 1 week previously).

The patient reported being treated with oral as well as long-acting risperidone injection during his previous hospital stay, along with sertraline and trazodone. These were confirmed in the medical records obtained from another facility.

Mr. B admitted to smoking a pack of cigarettes daily, drinking 12 beers a day, smoking a bag of marijuana daily, and using cocaine occasionally. Mr. B said that he lived by himself in an apartment. He did not work and received financial assistance from the state for his disability, i.e., schizophrenia and bipolar disorder.

From previous hospitalization records, the patient had been admitted to this hospital with complaints of hearing voices that told him to kill himself in the context of recent use of alcohol, cocaine, and marijuana 2 months before this presentation. The next day, when it was discovered that a female patient on the unit was related to him, the patient was told that he had to be transferred to another hospital. At this point, the patient became upset and said that he was not hearing voices anymore, nor was he feeling suicidal or homicidal and did not want to go to another hospital.

Mr. B said that he was no longer hearing voices, but he continued to feel suicidal as well as homicidal (against his sister). He denied having any means with which he could kill her, and this was the only factor that prevented him from killing her. When questioned regarding the possible consequences of killing his sister, Mr. B responded that he was confident that he would not go to jail because his entire family would support him in his act. When gently confronted with the idea that his family support may have little to do with the legal accountability of his behavior, he became irritable, and the conversation could not be pursued further. His affect remained flat throughout this interview.

On the second day of hospitalization, Mr. B was seen to be out and about on the unit, interacting with peers and staff members and attending all group therapy sessions, but continuing to express homicidal ideation.

On the third day, the patient's homicide risk was assessed using the method suggested by Borum and Reddy [2]. (1) Attitudes: The patient did not have any history of prior violence. His family history did not support violence. He continued to pray. (2) Capacity: He denied having considered a method of homicide, so capacity could not be adequately assessed. He said, however, that he was going to move to Florida (from the Midwest) to remove himself from his sister, which would decrease opportunity/capacity. (3) Threats: Although Mr. B had complained to his sister and her coworker in the past about her stealing the money from their inheritance, he denied having threatened her, having been physically aggressive towards her, or having done anything to prepare to harm her. (4) Intent: He denied current intent. (5) Others: He said that his family may support him if he killed her, as they were also angry at her for stealing, but he did not feel that they would join him or encourage him in doing so. (6) Noncompliance: He appeared to be compliant with risk reduction measures. Additionally, he denied access to a weapon. He provided his physician verbal permission to contact his sister. He finally said that he did not want to kill her. From this assessment, the final impression was that the risk of violence was low and that his physicians did not have sufficient reason to either notify police or pursue involuntary hospitalization. After his sister was contacted and notified by the resident about his expressed feelings and violent thoughts, Mr. B was discharged from the hospital to enter a substance abuse inpatient rehabilitation facility.

After the patient's discharge, some of his belongings were discovered in the possession of the female patient on the unit. When she was confronted, she confessed that she was Mr. B's "significant other" (with a different last name) and they had arrived in the ER together but had not disclosed their relationship (as they had not been allowed to stay together on the same unit during their previous visit to this hospital). Moreover, they had decided to simultaneously feign similar psychiatric symptoms of hearing voices and feeling suicidal and homicidal to gain admission because they were homeless and needed a place to stay for a week. This confession provided a rare confirmation of the suspicion that the symptoms of homicidality were feigned for external incentive. In retrospect, the Borum factors proved to be helpful in accurately evaluating low risk of harm in this case.

Homicide risk assessment supported feigned homicidality based upon the following. (1) The patient was known to feign similar symptoms for a clear external incentive (housing) during the previous two visits; (2) he was currently experiencing an acute need for housing; (3) he had not been compliant with his medications for several months but had not "relapsed" until 3 days previously; (4) Mr. B denied knowing the female patient, with whom he had presented to the same ER during his past two visits; (5) the patient did not have easy access to a weapon; (6) the patient had planned to move to Florida after 2 weeks; and (7) he was not experiencing acute psychotic symptoms. The homicide risk was assessed to be low, and the patient was discharged from the ER after medical stabilization for hypertension.

3. Discussion

The cases highlight several important clinical points towards the assessment of feigned homicidality. Keeping these two cases in mind, we would now like to discuss how to differentiate feigned homicidality from pseudologia fantastica, how to evaluate a verbal threat, how to assess homicidal risk, and eventually how to conduct further assessment when feigned homicidality is suspected We then discuss the duty to protect and the use of a multidisciplinary approach in the management of such cases.

3.1. Differentiating Feigned Homicidality from Pseudologia Fantastica

Today, pseudologia fantastica is not to be found in the DSM 5 [3], although factitious disorder is, earlier known as Munchausen's Syndrome. The original story of Van Munchausen was of a legendary baron who told fanciful false episodes of adventure. Kraepelin referred to Munchausen when describing psychological processes of "hunter traits", which he considered within the realm of normalcy. Delbruck [4] was the first to use the term "pseudologia fantastica" in 1891 and describe it as a phenomenon that is observed in normal, characterologically disturbed, as well as mentally ill individuals. It exists on a continuum between lying with full awareness of the statement's untruths and delusions where the unreality of the belief is not appreciated. In pseudologia fantastica, the subject's vulnerability to self-deceit from his own falsehoods can vary and fluctuate.

In summarizing the features of pseudologia fantastica from the literature, Birch and colleagues [5] list "excessive, impulsive lying" that usually begins in adolescence and then persists. The lies are fanciful, easily shown to be falsehoods, and are self-destructive. Motives are internal such as wishful fantasy, and not externally motivated, such as to gain materially or avoid punishment. Pseudologia fantastica involves impaired reality adherence but not to the degree found in delusions. More recently, Thom et al. [6] described the phenomenon of pseudologia fantastica as a state of chronic lying/storytelling that is dramatic, seeking admiration or sympathy, and out of proportion to the obvious benefit.

3.2. Evaluating the Verbal Threat

Regardless of whether a homicidal threat is initially thought to be idle or serious, feigned or authentic, it should be assessed. Not many years after the California Supreme Court's Tarasoff ruling in 1976 [7], Appelbaum [8] provided a three-step approach for dealing with homicidal threats that remains today as prudent guidance. Although the screening is derived from early Tarasoff-like case laws, this three-step approach seems self-evident in its practicality for any clinical risk assessment including the risk of suicide assessment: (1) Gather data that are relevant to the assessment of dangerousness. (2) Based upon this data, make a determination of dangerousness and select a course of action and (3) implement this plan. By case examples, Appelbaum illustrates how clinicians overreact and issue unnecessary warnings following insufficient assessment of the threat. Here, it is suggested additionally that assessment informs the action to be taken not only by a methodical determination of the patient's level of seriousness and dynamics of the risk, but also by determining whether the threat is authentic or feigned. Whether a verbal threat of homicide is feigned is also relevant to its seriousness.

In theory, as with risk assessments in general, probability predictions that a homicidal threat will be carried out should be more comprehensive than just to answer forced, dichotomous questions such as whether to hospitalize or whether to issue protective warnings [9]. To address such protective actions to be taken, the assessment of the risk, whether associated with a verbal threat or not, should incorporate four inquiries [8]: (1) Is the patient dangerous to others? (2) Is the danger due to serious mental illness? (3) Is the danger imminent? and (4) Are potential victims of the danger reasonably identifiable? The present discussion focuses only on those patients who would be hospitalized either because the danger is due to serious mental illness in need of intensive treatment, or the danger is imminent and no less restrictive means of protection are available.

Whenever practical, it is helpful to explain to the patient at the beginning of the clinician–patient relationship the terms of confidentiality, including the possible exception of protective warnings. Assessment of the potential for externally directed violence should then proceed along with the assessment of the potential for self-harm and suicide. Family members, previous treatment providers, and other collateral sources should be contacted where doubt exists about the seriousness of an expressed threat [6]. As with the assessment for suicide, any initial threat of homicide is taken seriously and evaluated accordingly, even when the clinician suspects that the threat is feigned.

3.3. Assessment of Risk

With the realization that research on assessing the seriousness of homicidal threats is much less developed than research on the risk of aggressive behavior in general, the clinician can usefully apply the guidelines provided by Borum and Reddy [2] for assessing the seriousness of the risk of homicide. Practical, methodical, and literature-based, the Borum and Reddy factors are summarized with the acronym ACTION: (1) Does the patient hold Attitudes that support violence as a method to deal with the interpersonal conflict? (2) Does he have the Capacity to carry out the threat? There are several components to capacity: availability of the intended victim, availability of means (e.g., weapon), and physical capacity as well as mental capacity. (3) Has he already crossed Thresholds in the direction of violating the victim's autonomy and privacy or of preparing to act violently against the victim (e.g., stalking, verbal threats, inappropriately accosting the victim at the victim's place of employment)? (4) Does the patient have a serious Intent behind the threat? Is it conditional and will only be carried under specific conditions that are not yet in place? (5) Do the attitudes and reactions of significant Others support violent responses? Conversely, do family members and close friends favor non-violent approaches to interpersonal problems? (6) And finally, is the patient Non-compliant with attempts at risk reduction, such as following treatment recommendations, helping to identify and contact the victim (if a protective warning is indicated), and neutralization of firearms. The reverse, or absence of the Borum factors, can be considered as potentially protective against enactment of the threat. Ultimately, the determination of high, medium, and low risk for violence following a verbal threat of homicide is a matter of clinical judgment, and not numerically quantitative.

3.4. Assessment of Feigned Homicidality

If the suspicion of feigned homicidality persists even after the homicidal threat has been assessed, it should be further explored. Psychiatrists need more than just a gut feeling to determine if the patient is lying. Psychiatrists must be aware of their countertransference while conducting further assessment. Guidelines for malingering can be useful in this regard, with malingering being "the intentional production of false or grossly exaggerated physical or psychological symptoms, motivated by external incentives such as avoiding military duty, avoiding work, obtaining financial compensation, evading criminal prosecution or obtaining drugs" [3]. The clinician is challenged with a dearth of professional literature on the assessment of feigned homicidality and by the patient's disinclination to explicitly share his motivation for malingering, which would of course defeat the purpose.

Methods for the assessment of malingering in general should be useful in detecting feigned homicidality [1]: namely, inconsistencies (1) in the patient's self-reports of homicidal thoughts (threats); (2) between what the patient reports and what others observe and report; (3) in observations concerning the "symptoms" themselves, in this case the patient's potential for homicide; (4) between the patients' self-report and performance on psychological tests (or mental status exam); and (5) in the patient's report of homicidal threats or thoughts and how actual threats with serious intent are manifested. The task is challenged by the fact that the patient only needs to express homicidal ideations to arouse concern; he need not necessarily malinger a complex condition such as psychosis.

The guidelines proposed by Borum and Reddy [2] are useful in the assessment of risk of homicidality. As in the case of Mr. B, they proved useful in highlighting the discrepancies in his history and the assessment of the overall risk. Moreover, due to safety concerns, it is often difficult to entertain the possibility of feigning symptoms when the risk of homicidality is assessed to be high. However, when the risk is determined to be low, the third dimension, i.e., feigning of homicidality can be further explored, as highlighted in the case of Mr. B. Although it is not always possible to gain knowledge of the potential incentives, an active search may prove to be useful. The areas that may assist in the determination of malingering include (1) a thorough history from the patient—it may even be helpful to periodically reassess the patient and be cognizant of any discrepancies in the narration over time; (2) thorough collateral information—this can prove to be very helpful, especially when obtained from multiple sources; refusal to provide any collateral source and lack of cooperation during the diagnostic or treatment process may also point towards an ulterior motive; (3) an active inquiry regarding the pertinent motives—financial constraints, lack of housing, and legal issues are especially prevalent in certain areas; and (4) legal inquiry—certain states provide an easy to access legal history of their clients, which may disclose any legal matter that the patient might try to avoid or delay by staying in the hospital.

3.5. *A Multi-Disciplinary Approach*

Cases with possible legal implications should involve a multidisciplinary approach to minimize risk and improve quality of assessment and management. Obtaining health records from other medical facilities, reviewing prior medical records, and obtaining collateral information from family members can yield vital sources of information in such cases. Ruling out possible medical issues or lies, such as was done in the Case 1 example above, can provide useful pieces of information in support of malingering. Medical providers may need to involve the ethics committee of the hospital, or consult with the legal department if they are unsure about addressing the legal aspects in these cases, such as in the context of duty to warn, and possible implications of discharging such a patient from the emergency room.

3.6. *Duty to Protect*

In the landmark ruling of Tarasoff v. The Regents of University of California [7], the court ruled that therapists need to take reasonable actions to protect potential victims of possible harm from their dangerous patients. When a patient makes a verbal threat against an identifiable victim, and the provider assesses such threat to be credible, the provider is directed to warn potential victims, inform the authorities, and/or take other necessary action to protect the life of potential victims [10]. Different states in the U.S. have adopted different standards in this area. While some states recommend a duty to warn potential victims, other states recommend a duty to take all necessary steps in addition to warning potential victims, to save lives. Some U.S. states mandate while others merely permit a breach of patient–provider confidentiality to carry out the established duty to warn/protect. While duty to warn is an accepted exception to confidentiality in most countries, only recently has it gained a legal precedent or background in other developed countries. [11]

4. Conclusions

We summarize the recommended steps for the evaluation and management of feigned homicidality in Table 2.

With the use of two case examples, we highlighted the assessment and management of feigned homicidality with an emphasis on the risk of verbal threat, assessment of the seriousness of the risk, differentiation from pseudologia fantastica, understanding duty to protect, and using a multidisciplinary approach.

We propose the addition of the assessment of malingering to the risk assessment for homicidality, especially in cases where there are enough reasons to suspect it. For example, a thorough screening for inconsistencies in the narration and presentation as well as potential external incentives for feigning homicidal symptoms may be included while using Borum and Reddy factors (expanding the acronym to ACTION). Other scales may be developed to include this dimension of homicidality.

Table 2. Steps recommended towards the evaluation and management of feigned homicidality.

Inquiries for the evaluation of the verbal threat	Is the patient dangerous to others? Is the danger due to serious mental illness? Is the danger imminent? Are potential victims reasonably identifiable?
Assessing the seriousness of the risk of homicide (ACTION)	Attitude that supports violence Capacity to carry out the threat Threshold crossed towards violence Intent is serious Others in life support violent responses Non-compliant with risk reduction methods
Differentiation from pseudologia fantastica	Pseudologia fantastica manifests as excessive and impulsive lying that is fanciful, done to seek attention or sympathy, and out of proportion to the motive
Assessment of malingering (feigning) when suspected	Locate inconsistencies in the narrative using a. thorough history b. Thorough collateral information c. Active inquiry regarding pertinent motives d. Legal inquiry
Multidisciplinary approach	Consider consultation with ethics committees, legal counsel, and other specialists as warranted.
Duty to Protect	Take reasonable steps as recommended by the guidelines in that specific jurisdiction

Author Contributions: A.R.F. conceptualized and designed the study. All authors are psychiatrists who collectively conducted the clinical interviews, wrote the manuscript, and contributed to editing the paper. All authors revised and reached an agreement on the final version of the work. All authors have read and agreed to the published version of the manuscript.

Funding: No funding was obtained for this project.

Institutional Review Board Statement: As a policy, the institution's IRB does not review articles involving a case report or case series with two cases.

Informed Consent Statement: Written informed consent for publication was obtained from participating patients.

Data Availability Statement: Not applicable.

Conflicts of Interest: The authors declare no conflict of interest.

References

1. Resnick, P.J. *Malingering, in Principles and Practice of Forensic Psychiatry*; Rosner, R., Ed.; Arnold: London, UK, 2003; pp. 543–554.
2. Borum, R.; Reddy, M. Assessing violence risk in Tarasoff situations: A fact-based model of inquiry. *Behav. Sci. Law* **2001**, *19*, 375–385. [CrossRef]
3. American Psychiatric Association. *Diagnostic and Statistical Manual of Mental Disorders*, 5th ed.; American Psychiatric Association: Washington, DC, USA, 2013.
4. Delbruck, A.W.A. *Die Pathologische Luge und Die Psychisch Abnormen Schwindter*; Verlag Von Ferdinand Enke: Stuttgart, Germany, 1891.
5. Birch, C.D.; Kelln, B.R.C.; Aquino, E.P.B. A review and case report of pseudologia fantastica. *J. Forensic Psychiatry Psychol.* **2006**, *17*, 299–320. [CrossRef]
6. Thom, R.; Teslyar, P.; Friedman, R. Pseudologia Fantastica in the Emergency Department: A Case Report and Review of the Literature. *Case Rep. Psychiatry* **2017**, *2017*, 1–5. [CrossRef] [PubMed]

7. Supreme Court of California. Tarasoff v. Regents of the University of California. 1976. Available online: https://scholar.google.com/scholar_case?case=263231934673470561 (accessed on 20 December 2021).
8. Appelbaum, P.S. Tarasoff and the clinician: Problems in fulfilling the duty to protect. *Am. J. Psychiatry* **1985**, *142*, 425–429. [PubMed]
9. Felthous, A.R. The clinician's duty to protect third parties. *Psychiatr. Clin. N. Am.* **1999**, *22*, 49–60. [CrossRef]
10. Cooper, A.E. Duty to Warn Third Parties. *JAMA* **1982**, *248*, 431–432. [CrossRef] [PubMed]
11. Gavaghan, C. A Tarasoff for Europe? A European Human Rights perspective on the duty to protect. *Int. J. Law Psychiatry* **2007**, *30*, 255–267. [CrossRef] [PubMed]

Article

Forensic Application of Monoclonal Anti-Human Glycophorin A Antibody in Samples from Decomposed Bodies to Establish Vitality of the Injuries. A Preliminary Experimental Study

Benedetta Baldari [1,†], Simona Vittorio [1,†], Francesco Sessa [2], Luigi Cipolloni [2,*], Giuseppe Bertozzi [2], Margherita Neri [3], Santina Cantatore [2], Vittorio Fineschi [1] and Mariarosaria Aromatario [4]

1 Department of Anatomical, Histological, Forensic and Orthopedic Sciences, Sapienza University of Rome, 00161 Rome, Italy; benedetta.baldari@uniroma1.it (B.B.); simonavittorio1@gmail.com (S.V.); vittorio.fineschi@uniroma1.it (V.F.)
2 Section of Legal Medicine, Department of Clinical and Experimental Medicine, University of Foggia, 71122 Foggia, Italy; francesco.sessa@unifg.it (F.S.); giuseppe.bertozzi@unifg.it (G.B.); santina.cantatore@unifg.it (S.C.)
3 Department of Medical Sciences, University of Ferrara, 44121 Ferrara, Italy; margherita.neri@unife.it
4 Legal Medicine Division, Sant'Andrea University Hospital, 00189 Rome, Italy; maromatario@ospedalesantandrea.it
* Correspondence: luigi.cipolloni@unifg.it; Tel.: +39-0881-736926
† These authors shared first authorship.

Citation: Baldari, B.; Vittorio, S.; Sessa, F.; Cipolloni, L.; Bertozzi, G.; Neri, M.; Cantatore, S.; Fineschi, V.; Aromatario, M. Forensic Application of Monoclonal Anti-Human Glycophorin A Antibody in Samples from Decomposed Bodies to Establish Vitality of the Injuries. A Preliminary Experimental Study. *Healthcare* **2021**, 9, 514. https://doi.org/10.3390/healthcare9050514

Academic Editor: Mariano Cingolani

Received: 23 March 2021
Accepted: 20 April 2021
Published: 29 April 2021

Abstract: Glycophorins are an important group of red blood cell (RBC) transmembrane proteins. Monoclonal antibodies against GPA are employed in immunohistochemical staining during post-mortem examination: Through this method, it is possible to point out the RBC presence in tissues. This experimental study aims to investigate anti-GPA immunohistochemical staining in order to evaluate the vitality of the lesion from corpses in different decomposition state. Six cases were selected, analyzing autopsies' documentation performed by the Institute of Legal Medicine of Rome in 2010–2018: four samples of fractured bones and three samples of soft tissues. For the control case, the fracture region of the femur was sampled. The results of the present study confirm the preliminary results of other studies, remarking the importance of the GPA immunohistochemical staining to highlight signs of survival. Moreover, this study suggests that the use of this technique should be routinely applied in cases of corpses with advanced putrefaction phenomena, even when the radiological investigation is performed, the macroscopic investigation is inconclusive, the H&E staining is not reliable. This experimental application demonstrated that the use of monoclonal antibody anti-human GPA on bone fractures and soft tissues could be important to verify whether the lesion is vital or not.

Keywords: forensic pathology; glycophorin A investigation; vitality

1. Introduction

Glycophorin (GP) A, GPB, GPC, and GPD are an important group of red blood cell (RBC) transmembrane proteins, described for the first time by Fairbanks et al. [1]. The four varieties of glycophorin represent approximately 2% of the total red blood cells (RBC) membrane protein: among the others, the predominant RBC glycophorin is GPA (glycophorin A—0.5 million copies per RBC). These proteins are characterized by a high content of sialic acid, giving a negative surface charge to the RBC [2]. For this reason, glycophorins play a pivotal role in controlling several essential functions such as the interactions both with vascular endothelium and with other blood cells. Moreover, GPA and B characterize the antigenic determinants, respectively, for the MN and Ss blood groups. Although the molecular function of GPA remains not fully understood, several studies demonstrate that it interacts with band 3, the significant erythrocytes membrane-spanning protein [3].

On the other hand, as red cell antigens, GPA can act as a binding site for specific antibodies. Monoclonal antibodies against glycophorin A (GPA), indeed, are frequently used in immunophenotypic studies for the identification of erythroid precursors in hematologic disorders. In the forensic field, the anti-human glycophorin A (GPA) monoclonal antibodies are commonly used in human blood detection, identifying red blood cell membrane antigen [4]. Moreover, monoclonal antibodies anti-GPA can be used in immunohistochemical staining at post-mortem examination to point out RBC presence in tissues [5]. Thus, if in non-putrefied corpses, the macroscopic evidence of hemorrhagic tissue infiltration is commonly considered a macroscopic sign of the viability of a lesion, research using anti-GPA antibodies could allow the differential diagnosis between ante-mortem and post-mortem lesions. This differentiation, indeed, is just as important as the identification of the exact cause of death, remaining the most notable challenge for the forensic pathologist. In this scenario, the definition of the vitality of skin and bone injuries is inscribed. For example, from a macroscopic perspective, it is well-known that the red-purplish coloration of a cut or bruise in fresh skin is indicative of its vitality. In more difficult cases, the microscopic investigation can be performed even just to confirm the macroscopic data [6]. Contrariwise, in the case of decomposed corpses, histological and immunohistochemical investigation can be considered mandatory: the transformative phenomena could mask or conceal the signs of traumatic injuries [7–10]. As previously described, a large number of immunohistochemical staining is frequently used in order to establish the vitality in skin lesions, such as anti-α-1 chymotrypsin, anti-fibronectin, anti-TGF-α and TGF-ß1, anti-inflammatory cytokines, anti-TNFα, anti-adhesion molecules, and anti-tryptase [11–19]. However, these antibodies show reactivity on well-preserved tissues, while their use remains uncertain when the post -mortem examination is performed in decomposed corpses or in human skeletal remains. Nevertheless, in the forensic field, the application on human bones of the immunohistochemical staining through the GPA antibody is not completely investigated.

In this setting, this experimental study aims to investigate anti-GPA antibody immunohistochemical staining in order to evaluate the vitality both of soft tissue injuries and bone fractures sampled from corpses in advanced decomposition state. Moreover, a comparison, between the hematoxylin-eosin (H&E) staining and anti-human GPA antibody immunohistochemistry technique, is also provided in the present manuscript to evaluate their potential in identifying "vital" processes in putrefied soft-tissue-free bone and in soft tissues.

2. Materials and Methods

All cases were collected among the documentation of all autopsies performed by the Institute of Legal Medicine of Rome from 2010 to 2018 (about 1500 autopsies). Six cases were selected on the following criteria: unnatural death, traumatic blunt injuries, body conservation status (almost chromatic phase of putrefaction and no macrofauna-related lesions), estimated post-mortem interval. Three cases were selected as control cases, dealing with a subject who died from drowning with a femur fracture occurred in the post-mortem period, during the translation operations, and two experimentally produced injuries (skin and soft tissues surrounding laryngeal area). The main characteristics of each case were summarized in Table 1.

Table 1. The main characteristics of the selected case. PMCT: Post-mortem computed tomography; PMI: Post-mortem interval; CTL: Control.

Case (Tissue)	Sex and Age	Type of Offence	PMCT	State	Estimated PMI
1 (cranial)	F 57 year	Homicide	Yes	Later stage of Putrefaction	65 days
2 (vertebral)	M 30 year	Accident	No	Later stage of Putrefaction	70 days
3 (mandible)	F 39 year	Homicide	No	Early stage of Putrefaction	2 days
4 (larynx)	F 46 year	Homicide	No	Later stage of Putrefaction	181 days
5 (neck and rib)	F 37 year	Homicide	Yes	Later stage of Putrefaction	85 days
6 (retina)	F 3 months	Homicide	No	Early stage of Putrefaction	7 days
CTL 1	F 9 year	Accidental	Yes	Later stage of Putrefaction	187 days
CTL 2	M 29 year	Experimental	Yes	Early stage of Putrefaction	15 days
CTL 3	M 52 year	Experimental	No	Later stage of Putrefaction	45 days

For the purposes of this experimental study, the following tissues were sampled.

Case 1 dealt with a 57-year-old woman who died after a failed tentative of sexual assault. Several fractures were found affecting the cranial region. The external marginal area of the occipital fracture was sampled.

Case 2 concerned a 30-year-old male who died from other than traumatic causes. During the autopsy, a complete fracture of vertebral D4 was detected, whose sample was taken.

Case 3 was about a 39-year-old woman who died from acute cardio-respiratory insufficiency due to induced asphyxiation. Different injuries were found on the head. For the present study, a mandible fragment was sampled in the correspondence of the fracture.

Case 4 described the case of a 46-year-old woman who died from acute cardio-respiratory insufficiency from strangulation. Soft tissue surrounding laryngeal fracture was sampled.

Case 5 regarded a 37-year-old woman who died for homicidal violent mechanical asphyxiation; for this study, two tissues were collected: one sample from the II right rib, which was found fractured, and the other one from the neck skin, where signs of external compression were found.

Case 6 described the case of a three-month-old female, who died from cardiovascular arrest subsequent to massive brain edema, following a head injury. Considered the presence of bilateral retinal hemorrhages, the right retina tissue was sampled for the purposes of this study.

Finally, three cases were enrolled as negative controls: the case of a nine-year-old female, who died from drowning (Control 1). The corpse was exhumed six months after the burial. Before the autopsy a PMCT was performed, resulting negative for any injury. However, a femur-displaced fracture occurred in the post-mortem period, during the translation operations to the autopsy room. The lesions used as negative control of neck skin (Control 2) and tracheal soft tissues (Control 3) were experimentally produced.

At the end of the case selection, the following tissues were analyzed: four samples of fractured bones (cranium, vertebra, mandible, and rib) and three samples of different soft tissues (soft tissue of larynx, neck skin, and retina). Moreover, the fractured region of the femur (Control 1) was used as negative control case of bone injuries, while neck skin and tracheal soft tissues were used as negative control of soft tissue lesions.

Score rank, used in the present study to evaluate the force of immunohistochemical staining with anti-body anti-GPA, lied in a range from "negative" (marked as "−") to "positive" (indicated with "+").

In all cases, local prosecutors ordered autopsies to be performed to clarify the exact cause of death.

For all selected cases, H&E staining and anti-human GPA antibody immunohistochemistry investigation were performed both to establish RBC presence as vitality sign. In the present study, these stainings were performed on bones and on soft tissues for cases number 1, 2, 3, and 5. In case 4, the soft tissue of larynx was tested, while in case 6, the staining was performed in the retina.

On all sample collected during autopsy, 4 μm-thick paraffin-embedded sections were cut and stained with H&E staining following the standard protocol [20]. Staining with H&E was qualitatively classified as "reliable" (++), "not reliable" (–), and "not univocal" (+-), based on RBC morphologic identifiability.

In addition, anti-human GPA antibody immunohistochemistry investigation was performed using antibodies anti-glycophorin A (Santa Cruz Biotechnology, Inc., Dallas, TX, USA). The paraffin sections were mounted on slides covered with 3-aminopropyltriethoxysilane (Fluka, Buchs, Switzerland). Pre-treatment was necessary to facilitate antigen retrieval and to increase membrane permeability to antibodies anti-glycophorin A boiling 0.25 M EDTA buffer, at 20 °C. The primary antibody was applied in 1:500 ratio for glycophorin A and incubated for 120 min at 20 °C. The detection system utilized was the LSAB+ kit (Dako, Copenhagen, Denmark), a refined avidin-biotin technique in which a biotinylated secondary antibody reacts with several peroxidase-conjugated streptavidin molecules. The sections were counterstained with Mayer's haematoxylin, dehydrated, cover slipped, and observed in a Leica DM4000B optical microscope (Leica, Cambridge, UK).

3. Results

In the present study, six cases were selected: five women and one male. In four cases, the corpses were found in advanced decomposition state (cases numbered 1, 2, 4, and 5). The histological investigations were very complex, considering the post-mortem transformative processes. Contrariwise, the corpses were in early stage of putrefaction, characterized by discoloration of the abdomen up to marbling phenomena in cases 3 and 6.

3.1. *Macroscopic Analysis*

The macroscopic analysis did not allow ascertaining the exact cause of death; for this reason, the microscopic investigation was performed. Concerning the cause of death, in three cases, the asphyxiation was defined as the manner of death (cases numbered 3, 4, and 5); in two cases (cases numbered 1 and 6), the cranial trauma was identified as the decisive damage. Finally, in case numbered 2, the victim died from acute cardio-respiratory distress; the death was not related to direct trauma.

3.2. *Hemorrhagic Infiltration*

Hemorrhagic infiltration was not identified at macroscopical analysis in the cases characterized by the transformative phenomena. Consequently, it was necessary to analyze and collect soft tissues and bone samples, performing the microscopic investigation. The results are summarized in Table 2.

Table 2. The main results of the histological and immunoistochemical examination.

Case	H&E	Glycophorin
1 (cranial)	No Reliable	Positive on cranial fracture
2 (vertebral)	No Reliable	Positive on vertebral fracture
3 (mandible)	Reliable	Positive on mandible fracture
4 (larynx)	Not Univocal	Positive on soft tissue of larynx
5 (neck and rib)	No Reliable	Positive on wrinkle neck and rib
6 (retina)	Reliable	Positive on retina
CTL (different tissues)	Reliable	Negative on post-mortem fracture

H&E investigations were performed in order to highlight the RBC presence microscopically (Figure 1). In cases, numbers 1, 2, 4, and 5, hemorrhagic infiltrations were not found. Results were influenced by the postmortem transformative processes of tissues and bones.

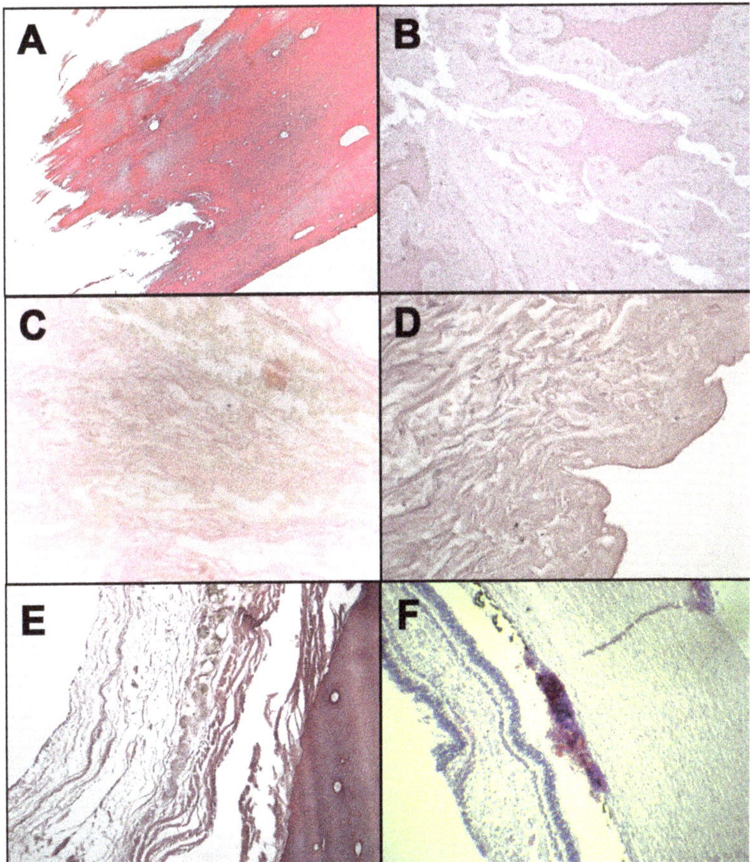

Figure 1. H&E staining: occipital bone from case 1 (**A**); mandible fragment from case 3 (**B**); soft tissues surrounding laryngeal fracture from case 4 (**C**); neck skin from case 5 (**D**); rib fracture from case 5 (**E**); retina from case 6 (**F**).

In cases 3 and 6, hemorrhagic infiltrations were macroscopically detected near the mandible fracture (case 3) and on the retinal tissue (case 6); moreover, these findings were also confirmed with the H&E analysis, highlighting the presence of RBC.

3.3. *Anti-Human GPA*

The anti-human GPA antibody immunohistochemistry investigation was performed in all cases in order to establish the RBC presence and to evaluate the sign of vitality in the detected lesions (Figure 2). Even if this technique confirmed the presence of the RBC in all sampled tissues, it was considered fundamental to determine the exact cause of death in four of the six cases selected.

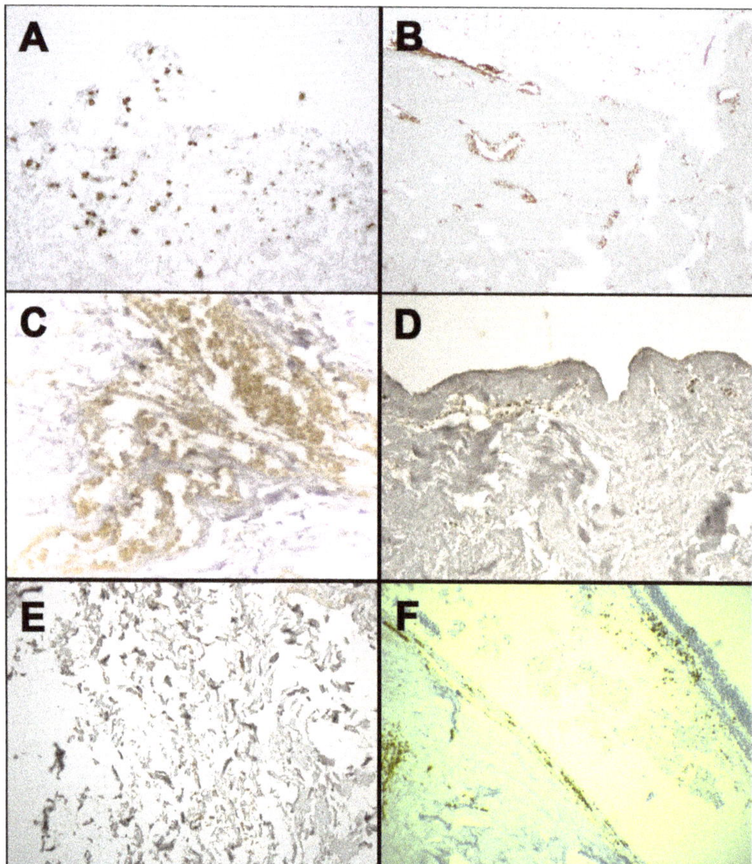

Figure 2. Immunohistochemical staining with anti-human GPA antibody: positivity at the margin of occipital bone fracture from case 1 (**A**); positivity at the stump of the mandible fracture from case 3 (**B**); positivity in the context of soft tissues surrounding laryngeal fracture from case 4 (**C**); positivity in dermis of neck skin from case 5 (**D**); positivity in the contest of soft tissues surrounding rib fracture from case 5 (**E**); positivity of the retinal sample from case 6 (**F**).

In case 1, the corpse was found in an advanced stage of decomposition and putrefaction. No radiological evidence of fracture was diagnosed; moreover, no macroscopical hemorrhagic infiltration was observed at pericranial soft tissue examination. However, a cranial linear compound fracture was found during the autopsy. H&E staining was not significant. Therefore, no evidence allows establishing if the fracture was caused by ante-mortem or post-mortem injury.

The GPA immunohistochemical investigation highlighted the presence of erythrocytes in the context of tissue cytoarchitecture, suggesting that the cranial fracture was vital, in accordance with the suspect's declaration.

In cases 2 and 4, the corpses were found with advanced putrefactive phenomena. In both cases, no PMCT was performed. A vertebral fracture was found in case 2; in case 4, a suspicious area near the soft tissue of larynx was sampled. No hemorrhagic infiltration was noticed macroscopically in the two corpses, no unique interpretation was concluded after H&E staining examination. For these reasons, the immunohistochemistry was performed, confirming the presence of RBC, as the vitality of the lesion.

In case 5, the post-mortem decomposition was detected. PMCT and H&E investigations were performed, even if no important evidence was collected in order to establish the exact cause of death. The anti-GPA antibody research on the neck skin sampled was performed in order to obtain data about the lesion, resulting positive. The presence of erythrocytes allows confirming that the injury occurred when the victim was alive.

Contrariwise, in cases 3 and 6, the corpses presented only early decomposition stages. No PMTC was performed. Hemorrhagic infiltration was detected both at macroscopical and microscopical (H&E staining) investigation. The immunohistochemical anti-GPA analysis was positive, supporting other evidence about the ante-mortem origin of the injury.

Finally, the control cases were performed on different tissues (Figure 3), produced during the translation operations to the autopsy room. No reaction to the GPA immunohistochemical analysis was described.

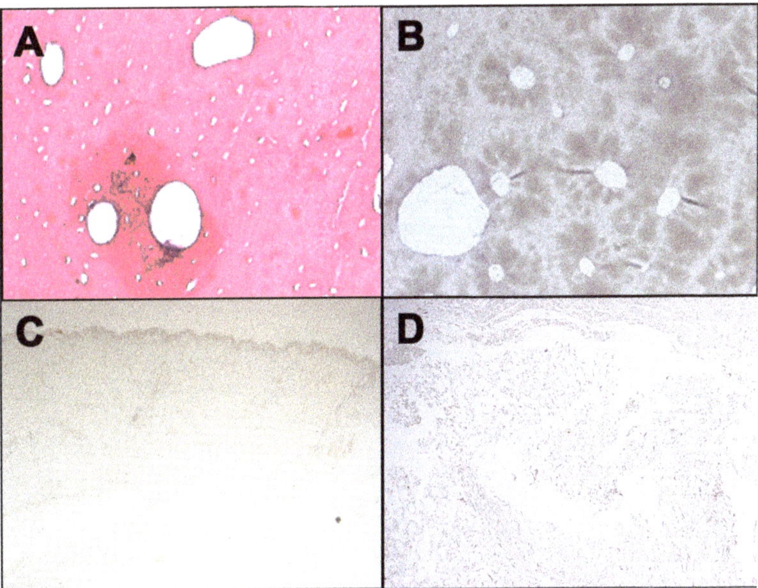

Figure 3. Negative immunohistochemical staining with anti-human GPA antibody for different tissues: Femur ((**A**), H&E; (**B**), immunostaining with anti-human GPA antibody); (**C**) Skin; (**D**) Soft tissues surrounding laryngeal area.

4. Discussion

Glycophorin is an integral membrane protein on the plasma membrane of erythrocytes. Monoclonal antibodies against glycophorin A are frequently used in clinical medicine for the identification of erythroid precursors in hematologic disorders [21]; moreover, this immunohistochemical staining can be used in forensic pathology for the identification of RBC in bone and tissue. According to the literature, to date, forensic application of the GPA immunohistochemical technique in order to evaluate the vitality of a wound is not quite investigated. As described by Cattaneo et al., the presence of clots and red blood cell residues on the fractured margins can be considered strongly indicative of vital reaction [6].

However, in this study, anti-glycophorin A analysis was performed on bones (cases number 1, 2 3, and 5) and on soft tissues sampled in corpses found in different postmortem period: This technique was applied on the soft tissue of larynx in case 4, on wrinkle neck in case 5, and on the retina in case 6.

Technical difficulties were described in handling decomposed bodies, considering the presence of artifactual alteration of tissue structure and microscopic features [22]. In the discussed study, hemorrhagic infiltration is not macroscopically evident in decomposed corpses (cases 1, 2, 4, and 5). Nevertheless, at the macroscopical examination, it was possible to observe the presence of hemorrhagic infiltration on the mandible fracture (case 3) and in the retinal tissue (case 6). It is essential to note that the bodies sampled in cases 3 and 6 were well preserved: In these cases, it was simple to detect hemorrhagic infiltration during autopsies.

Although it is likely that more information may be gleaned from fresh bodies in perfectly preserved states, decomposed bodies may reveal significant anatomical and pathological features that enable both the cause and manner of death to be established.

Even if it is well described that histopathology techniques can be used in the identification of the vitality of lesions in preserved tissue, the efficacy of these techniques in putrefied tissue is not demonstrated. In the present study, H&E investigation was not reliable in cases numbered 1, 2, 4, and 5: Considering that the corpses were recovered in the 5th stage of human decomposition process, it was impossible to define the presence of hemorrhagic infiltration. On the other hand, this technique was able to demonstrate the presence of RBC in the tissues sampled from cases number 3 and 6 (early stage of human decomposition process).

The immunohistochemical investigation was performed in all cases, showing interesting results. Monoclonal antibodies against GPA resulted positive in all analyzed cases, indicating the presence of RBC and demonstrating the vitality at the moment of the lesion. Indeed, the results in the control case were negative.

The discussed data confirmed that the histopathological investigation should be combined with the immunohistochemical examination: Indeed, evaluating the vitality of an injury, immunohistochemical diagnosis can provide reliable information [23]. Mainly, this study highlighted the importance of the GPA technique both on bones and on soft tissue in order to collect information on RBC presence, collecting information about the vitality of the lesion.

Bone fracture is described as a complete or incomplete disruption of bone tissue continuity: When it occurs during the life, the fracture triggers a regular tissue reaction [24–26]. Although fracture healing depends on the age of the individual and their nutritional status, age does not play an important role once adulthood has been reached [27]. The histological finding can demonstrate particular features in fractures following blast trauma [28]. Macroscopic morphological patterns of bone fracture are routinely used in forensic pathology and anthropology to distinguish between ante-mortem, peri-mortem, and postmortem injuries. Based on macroscopic and microscopic findings, it is possible to classify the fracture, avoiding inaccuracy [29]. Indeed, even if the presence of erythrocytes does not prove the vitality of the reaction, it could be used as a marker of bleeding to suggest another confirmative investigation on the vitality of injuries.

According to this experimental study, the glycophorin analysis is very important in corpses found in advanced decomposition state. Particularly, it could be very helpful when there is no evidence of hemorrhagic infiltrations, both at macroscopical analysis and histological investigation.

Even if in forensic pathology the use of the anti-GPA immunohistochemical staining remains controversial, according to the presented results, this immunohistochemical technique can be considered very useful both on bone structures and on soft tissues, particularly when the corpses have been found in an advanced state of decomposition. A negative result suggests the absence of erythrocytes in the sampled tissue: This means that the lesion could be related to post mortem trauma. On the contrary, a positive result can suggest the presence of erythrocytes in the tissue, ascertain that the lesion on bone and/or on soft tissue has been generated by pre or perimortem trauma. In a recent study, it was described the usability for forensic purposes of the immunochemical staining in corpses with different conservation status [30], as confirmed by this study. In addition, the presented results are in line with Taborelli at al.'s [5] research article on glycophorin A, as a good marker to orient diagnosis of vitality thanks to the high resistance at different PMI. Moreover, this study showed that this immunohistochemical investigation is able to keep its diagnostic reliability in different types of tissue. Therefore, this methodological approach should be routinely applied in corpses with advanced putrefaction phenomena, even when the radiological investigation is performed, the macroscopic investigation is inconclusive, and the H&E staining is not reliable. Agreeing with observation of Cappella et al. [31], this study confirms that searching for bleeding or tissue reactions in fragmented bones or in tissues with advanced taphonomy is useful to reach a diagnosis of ante-mortem or post-mortem injury. Considering the statement of the same authors, we argue that simultaneous application of histological and immuno-histochemical methods searching for inflammatory markers and hemorrhaging joined with research of bone reaction activity is useful to reach a sustainable diagnosis of vitality.

5. Conclusions

In conclusion, though certainly not conclusive, this experimental application demonstrated that the use of monoclonal antibody anti-human GPA on bone fractures as well as soft tissues could be important to verify whether the lesion is vital or not.

Author Contributions: Conceptualization, L.C., F.S., B.B. and G.B.; data curation, L.C., F.S., M.A. and S.C.; writing-original draft preparation, L.C., F.S., B.B., M.A. and S.V.; writing-review and editing, F.S., B.B. and S.V.; supervision, L.C., V.F. and M.N. All authors have read and agreed to the published version of the manuscript.

Funding: This research received no external funding.

Institutional Review Board Statement: Not applicable.

Informed Consent Statement: Not applicable.

Data Availability Statement: All data are included in the manuscript.

Conflicts of Interest: The authors declare no conflict of interest.

References

1. Fairbanks, G.; Steck, T.L.; Wallach, D.F.H. Electrophoretic analysis of the major polypeptides of the human erythrocyte membrane. *Biochemistry* **1971**, *10*, 2606–2617. [CrossRef] [PubMed]
2. Chasis, J.A.; Mohandas, N. Red blood cell glycophorins. *Blood* **1992**, *80*, 1869–1879. [CrossRef] [PubMed]
3. Giger, K.; Habib, I.; Ritchie, K.; Low, P.S. Diffusion of glycophorin A in human erythrocytes. *Biochim. Biophys. Acta BBA Biomembr.* **2016**, *1858*, 2839–2845. [CrossRef] [PubMed]
4. Turrina, S.; Filippini, G.; Atzei, R.; Zaglia, E.; De Leo, D. Validation studies of rapid stain identification-blood (RSID-blood) kit in forensic caseworks. *Forensic Sci. Int. Genet. Suppl. Ser.* **2008**, *1*, 74–75. [CrossRef]
5. Taborelli, A.; Andreola, S.; Di Giancamillo, A.; Gentile, G.; Domeneghini, C.; Grandi, M.; Cattaneo, C. The use of the an-ti-glycophorin a antibody in the detection of red blood cell residues in human soft tissue lesions decomposed in air and water: A pilot study. *Med. Sci. Law* **2011**, *51*, 16–19. [CrossRef] [PubMed]
6. Cattaneo, C.; Andreola, S.; Marinelli, E.; Poppa, P.; Porta, D.; Grandi, M. The detection of microscopic markers of hemor-rhaging and wound age on dry bone: A pilot study. *Am. J. Forensic. Med. Pathol.* **2010**, *31*, 22–26. [CrossRef] [PubMed]
7. Dreßler, J.; Bachmann, L.; Koch, R.; Müller, E. Enhanced expression of selectins in human skin wounds. *Int. J. Leg. Med.* **1998**, *112*, 39–44. [CrossRef]
8. Bonelli, A.; Bacci, S.; Vannelli, G.B.; Norelli, G.A. Immunohistochemical localization of mast cells as a tool for the discrimi-nation of vital and postmortem lesions. *Int. J. Legal. Med.* **2003**, *117*, 14–18. [CrossRef]
9. Casse, J.M.; Martrille, L.; Vignaud, J.M.; Gauchotte, G. Skin wounds vitality markers in forensic pathology: An updated review. *Med. Sci. Law* **2016**, *56*, 128–137. [CrossRef]

10. Sessa, F.; Varotto, E.; Salerno, M.; Vanin, S.; Bertozzi, G.; Galassi, F.M.; Maglietta, F.; Salerno, M.; Tuccia, F.; Pomara, C.; et al. First report of Heleomyzidae (Diptera) recovered from the inner cavity of an intact human femur. *J. Forensic Leg. Med.* **2019**, *66*, 4–7. [CrossRef]
11. Betz, P.; Nerlich, A.; Wilske, J.; Tubel, J.; Penning, R.; Eisenmerger, W. The immunohistochemical localization of al-pha1-antichymotrypsin and fibronectin and its meaning for the determination of the vitality of human skin wounds. *Int. J. Legal. Med.* **1993**, *105*, 223–227. [CrossRef]
12. Grellner, W.; Vieler, S.; Madea, B. Transforming growth factors (TGF-α and TGF-β1) in the determination of vitality and wound age: Immunohistochemical study on human skin wounds. *Forensic Sci. Int.* **2005**, *153*, 174–180. [CrossRef] [PubMed]
13. Grellner, W. Time-dependent immunohistochemical detection of proinflammatory cytokines (IL-1β, IL-6, TNF-α) in human skin wounds. *Forensic Sci. Int.* **2002**, *130*, 90–96. [CrossRef]
14. Dressler, J.; Bachmann, L.; Strejc, P.; Koch, R.; Müller, E. Expression of adhesion molecules in skin wounds: Diagnostic value in legal medicine. *Forensic Sci. Int.* **2000**, *113*, 173–176. [CrossRef]
15. Cipolloni, L.; Sessa, F.; Bertozzi, G.; Baldari, B.; Cantatore, S.; Testi, R.; D'Errico, S.; Di Mizio, G.; Asmundo, A.; Castorina, S.; et al. Preliminary Post-Mortem COVID-19 Evidence of Endothelial Injury and Factor VIII Hyperexpression. *Diagnostics* **2020**, *10*, 575. [CrossRef] [PubMed]
16. Fineschi, V.; Dell'Erba, A.S.; Di Paolo, M.; Procaccianti, P. Typical Homicide Ritual of the Italian Mafia (Incaprettamento). *Am. J. Forensic Med. Pathol.* **1998**, *19*, 87–92. [CrossRef]
17. Turillazzi, E.; Vacchiano, G.; Luna-Maldonado, A.; Neri, M.; Pomara, C.; Rabozzi, R.; Riezzo, I.; Fineschi, V. Tryptase, CD15 and IL-15 as reliable markers for the determination of soft and hard ligature marks vitality. *Histol. Histopathol.* **2010**, *25*, 1539–1546.
18. Pinchi, E.; Frati, P.; Aromatario, M.; Cipolloni, L.; Fabbri, M.; La Russa, R.; Maiese, A.; Neri, M.; Santurro, A.; Scopetti, M.; et al. miR-1, miR-499 and miR-208 are sensitive markers to diagnose sudden death due to early acute myocardial infarction. *J. Cell. Mol. Med.* **2019**, *23*, 6005–6016. [CrossRef]
19. Dettmeyer, R.B. *Forensic Histopathology: Fundamentals and Perspectives*; Springer International Publishing: Berlin/Heidelberg, Germany, 2018.
20. Salthouse, T.N. Histopathologic Technic and Practical Histochemistry. *Bull. Soc. Pharmacol. Environ. Pathol.* **1977**, *5*, 15. [CrossRef]
21. Sadahira, Y.; Kanzaki, A.; Wada, H.; Yawata, Y. Immunohistochemical identification of erythroid precursors in paraffin embedded bone marrow sections: Spectrin is a superior marker to glycophorin. *J. Clin. Pathol.* **1999**, *52*, 919–921. [CrossRef]
22. Byard, R.W.; Farrell, E.; Simpson, E. Diagnostic yield and characteristic features in a series of decomposed bodies subject to coronial autopsy. *Forensic Sci. Med. Pathol.* **2007**, *4*, 9–14. [CrossRef]
23. Dettmeyer, R.B. *Forensic Histopathology: Fundamentals and Perspectives: Vitality, Injury Age, Determination of Skin Wound Age, and Fracture Age BT*; Springer International Publishing: Cham, Germany, 2018; pp. 241–263.
24. Klotzbach, H.; Delling, G.; Richter, E.; Sperhake, J.; Püschel, K. Post-Mortem diagnosis and age estimation of infants' fractures. *Int. J. Legal. Med.* **2003**, *117*, 82–89. [CrossRef]
25. Bertozzi, G.; Sessa, F.; Maglietta, F.; Cipolloni, L.; Salerno, M.; Fiore, C.; Fortarezza, P.; Ricci, P.; Turillazzi, E.; Pomara, C. Immunodeficiency as a side effect of anabolic androgenic steroid abuse: A case of necrotizing myofasciitis. *Forensic Sci. Med. Pathol.* **2019**, *15*, 616–621. [CrossRef]
26. Borro, M.; Gentile, G.; Cipolloni, L.; Földes-Papp, Z.; Frati, P.; Santurro, A.; Lionetto, L.; Simmaco, M.; Fineschi, V. Personalised Healthcare: The DiMA clinical model. *Curr. Pharm. Biotechnol.* **2017**, *18*, 1. [CrossRef]
27. Weber, M.A.; Risdon, R.A.; Offiah, A.C.; Malone, M.; Sebire, N.J. Rib fractures identified at post-Mortem examination in sudden unexpected deaths in infancy (SUDI). *Forensic Sci. Int.* **2009**, *189*, 75–81. [CrossRef] [PubMed]
28. Pechníková, M.; Mazzarelli, D.; Poppa, P.; Gibelli, D.; Baggi, E.S.; Cattaneo, C. Microscopic Pattern of Bone Fractures as an Indicator of Blast Trauma: A Pilot Study. *J. Forensic Sci.* **2015**, *60*, 1140–1145. [CrossRef] [PubMed]
29. Delabarde, T.; Reynolds, M.; Decourcelle, M.; Pascaretti-Grizon, F.; Ludes, B. Skull fractures in forensic putre-fied/skeletonised cases: The challenge of estimating the post-Traumatic interval. *Morphologie* **2020**, *104*, 27–37. [CrossRef] [PubMed]
30. Lesnikova, I.; Schreckenbach, M.N.; Kristensen, M.P.; Papanikolaou, L.L.; Hamilton-Dutoit, S. Usability of Immunohisto-chemistry in Forensic Samples with Varying Decomposition. *Am. J. Forensic Med. Pathol.* **2018**, *39*, 185–191. [CrossRef] [PubMed]
31. Cappella, A.; Cattaneo, C. Exiting the limbo of perimortem trauma: A brief review of microscopic markers of hemorrhaging and early healing signs in bone. *Forensic Sci. Int.* **2019**, *302*, 109856. [CrossRef] [PubMed]

Case Report

A Rare Case of Suicide by Ingestion of Phorate: A Case Report and a Review of the Literature

Angelo Montana [1,*], Venerando Rapisarda [2], Massimiliano Esposito [1,*], Francesco Amico [1], Giuseppe Cocimano [1], Nunzio Di Nunno [3], Caterina Ledda [2] and Monica Salerno [1]

1 Legal Medicine, Department of Medical, Surgical and Advanced Technologies, "G.F. Ingrassia", University of Catania, 95123 Catania, Italy; francesco.amico05@community.unipa.it (F.A.); giuseppe.cocimano@you.unipa.it (G.C.); monica.salerno@unict.it (M.S.)

2 Department of Clinical and Experimental Medicine, University of Catania, 95123 Catania, Italy; vrapisarda@unict.it (V.R.); cledda@unict.it (C.L.)

3 Department of History, Society and Studies on Humanity, University of Salento, 73100 Lecce, Italy; nunzio.dinunno@unisalento.it

* Correspondence: angelo.montana@uniroma1.it (A.M.); massimiliano.esposito91@gmail.com (M.E.); Tel.: +39-3287655428 (A.M.); +39-3409348781 (M.E.)

Abstract: Phorate is a systemic organophosphorus pesticide (OP) that acts by inhibiting cholinesterases. Recent studies have reported that long-term low/moderate exposure to OP could be correlated with impaired cardiovascular and pulmonary function and other neurological effects. A 70-year-old farmer died after an intention ingestion of a granular powder mixed with water. He was employed on a farm for over 50 years producing fruit and vegetables, and for about 20 years, he had also applied pesticides. In the last 15 years, he used phorate predominantly. The Phorate concentration detected in gastric contents was 3.29 µg/mL. Chronic exposure to phorate is experimentally studied by histopathological changes observed in the kidney. In the light of current literature, our case confirms that there is an association between renal damage and chronic exposure to phorate in a subject exposed for years to the pesticide. Autopsies and toxicological analyses play a key role in the reconstruction of the dynamics, including the cause of the death.

Keywords: phorate; acute toxicity; chronic toxicity; histopathological kidney; toxicological examination

Citation: Montana, A.; Rapisarda, V.; Esposito, M.; Amico, F.; Cocimano, G.; Nunno, N.D.; Ledda, C.; Salerno, M. A Rare Case of Suicide by Ingestion of Phorate: A Case Report and a Review of the Literature. *Healthcare* **2021**, *9*, 131. https://doi.org/10.3390/healthcare9020131

Academic Editor: Christopher R. Cogle

Received: 5 January 2021
Accepted: 26 January 2021
Published: 29 January 2021

1. Introduction

Phorate (IUPAC name: O,O-diethyl S-[(ethylsulfanyl)methyl] phosphorodithioate; CAS Number: 298-02-2) is a systemic organophosphorus pesticide (OP) that acts by inhibiting cholinesterases, which are the enzymes involved in transmitting nerve impulses [1]. Phorate is highly toxic to birds, fish, and mammals (male rat oral LD50 for the metabolite phorate oxon = 0.88 mg/kg) [2], and accidental human exposure, resulting in death in some instances, has been reported [3].

The self-poisoning of organophosphorus pesticide is a major clinical and public health problem across many rural regions. Some studies have highlighted higher suicide rates among farmers than the general population [4–6]. Some reviews pointed out higher suicide rates among farmers than any other occupational group in the United Kingdom and in Australia [7–9]. In contrast, applicators in Italy had a lower rate of accidents and suicide [10]. In fact, Parrón et al. [11] showed a suicide rate in certain areas of Spain where there was greater use of phorate [12], and ecological and case studies suggested an association between organophosphate pesticide (OP) use and suicide [13].

The Agricultural Health Study (AHS) is a large, prospective cohort study of private pesticide applicators (mostly farmers), designed to study associations between cancer and other chronic diseases and farm-related exposures [14]. Pesticides are associated with the following cancer: lung cancer [15,16], pancreatic cancer [17], colon and rectal cancer [18,19] all lymphohematopoietic cancers [20] leukemia [21], Non-Hodgkin lymphoma [22], multiple myeloma [23], bladder cancer [24] brain cancer and melanoma [25,26].

In humans, chronic exposure to phorate has been shown to cause a reduced acetylcholinesterase (AChE) activity in both blood plasma and the brain [27]. Its mechanism of action consists in the inhibition of AChE activity by phosphorylating the serine hydroxyl group of the substrate-binding domain [28]. In cases where there is an accumulation of acetylcholine (ACh), the "cholinergic syndrome" can develop, which causes an overstimulation of nicotinic, muscarinic, and central ACh receptors; at the level of the central nervous system (CNS), this causes different effects such as headache, drowsiness, dizziness, confusion, blurred vision, slurred speech, ataxia, coma, and convulsions, until there is a block of the respiratory center [29–31]. Recent studies have highlighted that long-term low/moderate exposure to OP could be correlated with impaired neurobehavioral function or other neurological effects [32,33], while the studies have reported higher suicide rates among farmers who used it than in the general population [34,35]. Other investigations on the chronic toxicity of OP have shown kidney damage in animal models mediated by several mechanisms such as damage to the cell membrane and proteins through oxidative stress induced by the generation of free oxygen radicals, functional disturbances related to plasma membrane injury, cellular DNA damage, and activation of apoptosis-related p53 [35–43]. In the present study, we report a case of suicide carried out through the high ingestion of phorate of an agricultural worker exposed to phorate, at low doses, for several years. However, there is no evidence for an association between nephrotoxicity and chronic long-term exposure to low levels of phorate exposure in humans in the literature. Through this case report and a systematic review of animal and human models, this study aims to correlate kidney damage with chronic exposure to pesticide and phorate in the subjects exposed for years to the pesticide and dead after an intentional ingestion of phorate.

2. Materials and Methods

2.1. Case Description

A 70-year-old farmer was found disoriented and sweaty by his daughter. The man had been drinking a granular powder mixed with water. She immediately alerted the emergency services, but the man suddenly collapsed during transport to the emergency department. Despite resuscitation maneuvers, he died. He had been a farmer for over 50 years producing fruit and vegetables, and for about 20 years, he had used phorate. In the last nine years, he had used predominantly phorate twice/month, for 6–7 working h/day from March to July. From his medical records, compiled by his family doctor, it appears that the worker had had a history of depression for 10 years that had been treated pharmacologically with a specialist prescription.

No ethical committee was required. Written informed consent was obtained from the relatives.

2.2. Autopsy Findings

A complete autopsy was performed 48 h after death. External examination of the body showed a robust physique (height 165 cm, weight 81 kg), with hypostasis on the upper half of the body (head, neck, superior thorax, and upper limbs). The autopsy revealed the heart with a regular shape, weight 470 g, and measured $12 \times 9.5 \times 6.8$ cm; the coronary arteries were healthy with a right dominance. The myocardium and the valvular apparatus were normal. The analysis of the respiratory system was performed through the removal of the heart-lung block. The larynx was edematous; the lung parenchyma appeared increased in volume, heavy (left 430 g, right lung 500 g), the consistency was emphysematous during palpation with congestion and petechiae. The esophagus walls had a brownish coloration; the stomach contained 100 mL of brown liquid, which was sampled (see Figure 1a). Organs taken during the autopsy showed evident signs of diffused visceral congestion. The organ specimens were fixed in 10% buffered formalin and embedded in paraffin. Microscopy with hematoxylin-eosin staining was performed: the brain showed intraparenchymal haemorrhages, perineuronal and perivasal oedema. The lungs displayed subpleural and endoalveolar haemorrhages, massive endoalveolar edema, and scattered bronchopneumonia outbreaks; there was an infiltrate of peribronchiolar lymphocytes and neutrophils were detected in the lungs and laryngeal mucosa.

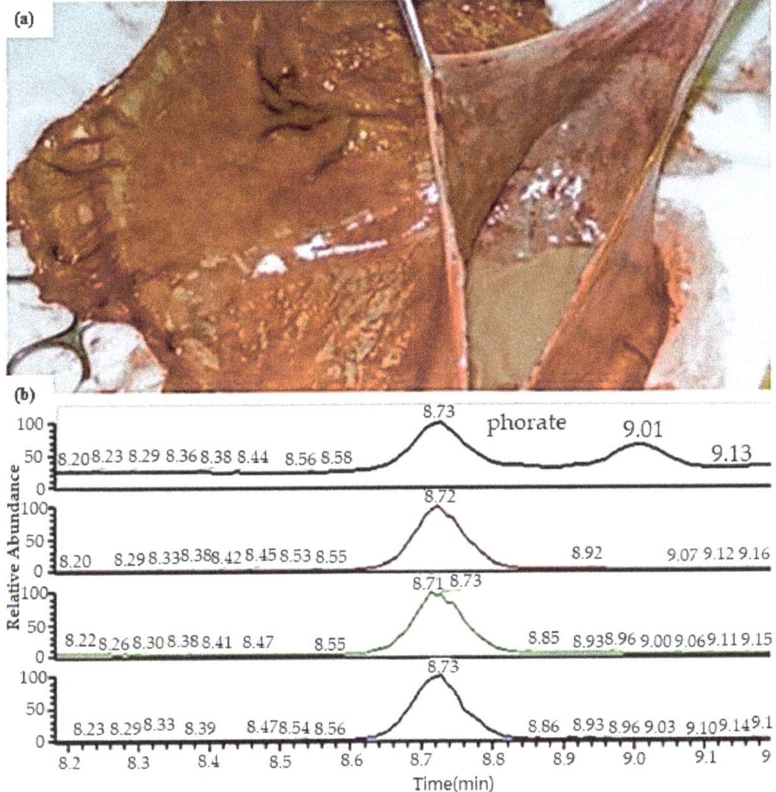

Figure 1. (**a**) the stomach's opening: stomach contained 100 mL of brown color; (**b**) chromatogram of the case.

Specimens of biological fluids (blood, urine, and gastric contents) taken during the autopsy were frozen (−20 °C). Sections of tissue were set up with hematoxylin and eosin for histological examination.

2.3. Poisoning

To determine what the subject had ingested, gastric contents were collected at the time of autopsy and stored at −20 °C, and 1 mL of samples was analyzed, after liquid/liquid extraction, by gas chromatography–mass spectrometry (GC–MS) analysis (single quadrupole).

2.4. Chronic Exposure Assessment

In the absence of chronic pre-mortem exposure data, the subject's kidney sections, after H&E staining, were observed to detect the presence of damage due to chronic exposure to phorate to verify if there had been an extended exposure during life.

2.5. Systematic Review

In accordance with the PRISMA statement a systematic review was performed [16].

2.6. Literature Search

SCOPUS, Medline (using PubMed as the search engine), Embase, and Web of Sciences databases were searched up to 30 September 2020 for the association of toxic phorate exposure with suicide and histopathological kidney changes in humans as the primary outcome.

MeSH was used with the following entry terms: "Phorate" AND "death"; "Phorate" AND "suicide"; "Phorate" AND "kidney". A search of the identified manuscripts was then made for inclusion suitability in this systematic review, and the research papers of significance were collected and reviewed.

2.7. Inclusion and Exclusion Criteria

The following criteria were used: (1) Studies that evaluated the correlation between Phorate and death (suicide, kidney damage, animals and humans). The following exclusion criteria were then used: (2) animal studies, (3) original articles in non-English language; and (4) editorials, posters, abstracts.

For duplicate studies, the article with detailed information was included.

2.8. Quality Assessment and Data Extraction

Two reviewers, A.M. and C.L. processed articles independently. The title, abstract, and full text of each potentially pertinent study was reviewed. Through the consultation and debate of additional reviewer V.R., any divergence on the eligibility of the studies was determined. The following information was extracted and organized from all suitable papers: authors, year of publication, the nationality of subjects, and study characteristics.

2.9. Characteristics of Eligible Studies

After a search of the scientific literature by reviewers, a total of 15 documents were collected.

Two were excluded after a subsequent review of the title and abstract, and 5 studies were ruled out after a review of the manuscript. In conclusion, 8 studies satisfied totally the inclusion criteria and were included in the systematic review. A flowchart depicting the choice of studies is shown in Figure 2.

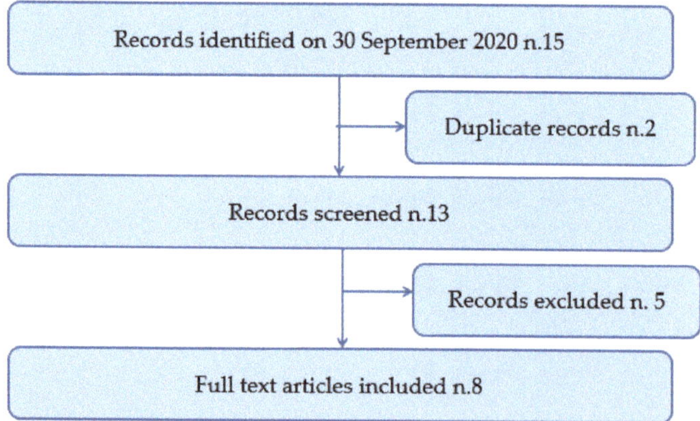

Figure 2. Flow diagram illustrating included and excluded studies in this systematic review.

A summary of the details of the included research papers is reported in Table 1.

Table 1. Characteristics of eligible studies (animals and humans).

Reference	Study Design	Target	Exposure	Intervention/Outcome	Main Findings
Khatiwada et al., 2012 [3]	case report	n.2 women: (1) 30-year-old (2) 80-year-old	acute	unintentional ingestion of phorate granules mistaken for food (sesame seeds)	30-year-old woman dead on arrival at the emergency department; 80-year-old woman survived after resuscitation procedures.
Peter et al., 2008 [44]	case report	30-year-old female	acute	impulsively swallowed 50 mL of phorate after a family dispute	survived after hospital treatment.
Qi et al., 2017 [45]	prospective, observational	rats; urine analysis	Chronic	rats were given a mixture of four op pesticides (dimethoate, acephate, dichlorvos, and phorate) for 90 days.	Alteration of kidney function, modification of DNAwith alteration of the metabolism of fatty acids, energy and sex hormones, antioxidant defense system.
Du et al., 2014 [46]	prospective, observational	rats; the plasma was analyzed	chronic	mixture of four op pesticides (dimethoate, acephate, dichlorvos, and phorate) for 24 weeks	kidney damage of tubular cell, granular and vascular degeneration.
Sun et al., 2012 [47]	prospective, observational	rats; metabonomics evaluation of urine by uplc-ms; long-term and low-level exposure	chronic	phorate daily in drinking water at low doses of 0.05, 0.15 or 0.45 mg/kg body weight (bw) for 24 weeks consecutively	kidney damage: the levels of creatinine (cr) and urea nitrogen (bun) were significantly elevated in the high-dose group, indicating kidney damage after exposure to phorate.
Li et al., 2016 [48]	prospective, observational	rats; the authors examined the effect of quercetina	chronic	mixture of four organophosphates (dichlorvos, acephate, dimethoate and phorate) for 90 days	kidney damage: histopathological examination showed extensive cell vacuolar denaturation and desquamation of the epithelial lining of the tubules; renal damage by impairing the reabsorption capacity of the proximal tubules and by decreasing glomerular filtration rate.
Mohssen 2001 [49]	prospective, observational	rats	chronic	subchronic inhalation of the recommended field dose of phorate (20 kg/ha).	kidney damage: impairment of glomerular function and tubular damage with mild to severe multifocal cloudy and hydropic degeneration (edema) with necrosis in kidney tubules.
Saquib et al., 2012 [50]	prospective, observational	rats	chronic	14 days of varying oral doses of phorate of 0.046, 0.092 or 0.184 mg	kidney damage: infiltration of leukocytes in the bowman's space with dilated blood vessels, and renal necrosis.

2.10. *Outcomes of Eligible Studies*

From the eight studies, only two were carried out on humans; both were case reports of acute exposure.

Peter et al. [44] reported a case of a 30-year-old woman who, after a family dispute, swallowed 50 mL of phorate. She survived after gastric lavage. Khatiwada et al. [3] reported a case of a 30-year-old woman and her 80-year-old grandmother who unintentionally ingested phorate granules mistaken for food (sesame seeds). The 30-year-old woman was pronounced dead on arrival at the emergency department; the 80-year-old woman survived after resuscitation procedures.

The other six studies were conducted on animals and related chronic exposure to phorate. The experiments were conducted on rats that were given only phorate or with other pesticides for a period from a minimum of 14 days to a maximum of 24 weeks consecutively. Due to chronic exposure, these rats showed kidney and DNA damage.

3. Results

In our case, phorate concentration detected in the gastric contents was 3.29 μg/mL (see Figure 1b).

According to the macroscopic and microscopic findings, the cause of death was attributed to respiratory failure with pulmonary dysfunction due to an acute cholinergic crisis. The larynx was edematous; the lung parenchyma appeared increased in volume, and the consistency was emphysematous during palpation. The esophagus walls had a brownish coloration; the stomach contained 100 mL of brown liquid, which was sampled. Histological examination of the lungs revealed a generalized stasis; it was observed that inside the lungs, the alveolar spaces were occupied by an eosinophilic proteinaceous material and some hemosiderin-laden macrophages associated with passive congestion. The use of a higher power microscope allowed us to see intra-alveolar edema and alveolar capillaries' engorgement.

The glomerulus exhibited hypercellularity (see Figure 3a). At certain places, leucocytic infiltration was noticed. There was a deposition of eosin-positive material between the tubules and degenerating tubules (see Figure 3b). A shrunken of the glomeruli were seen in some areas. Thus more spaces between Bowman's capsule and the glomerulus were displayed (see Figure 3c). There was increased cellularity in the glomerulus (see Figure 3d), tightly filling Bowman's capsule, and dilated tubules were seen with the separation of epithelial cells from the underlying basement membrane. The proximal and distal tubules showed hypertrophy of epithelial cells with obliterated lumina.

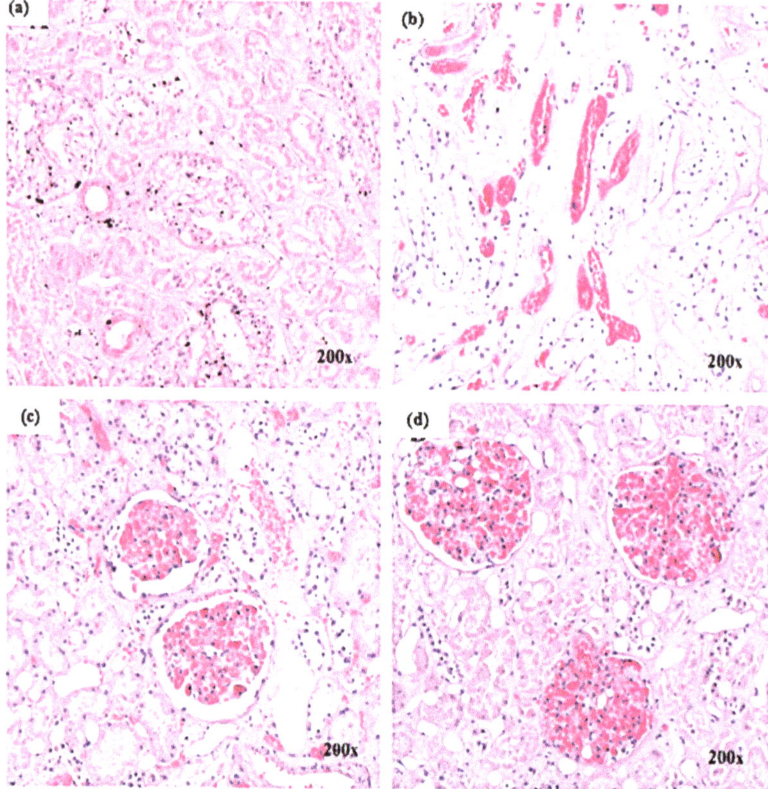

Figure 3. Histological findings in the kidney: (**a**) glomerulus hypercellularity; (**b**) deposition of eosin-positive material in the tubules; (**c**) increased space between Bowman's capsule and glomerulus; (**d**) increased cellularity in the glomerulus.

4. Discussion

In the case reported here, death followed fatal oral ingestion of phorate by a farmer who had used phorate for about 20 years. He had used predominantly phorate twice per month for 6–7 working h/day from March to July in the last nine years.

In the literature, there were never described fatal cases of accidental ingestion of phorate. It was described a case after unintentional ingestion of phorate by a 30-year-old woman in the literature. The woman confused phorate granules for sesame seeds, she mixed it with pickle and rice and consumed it. She was taken to the hospital and, after resuscitation, was transferred to the intensive care unit. She was extubated on the 17th day and discharged on the 23rd day [3].

In another case report a 28-year-old woman, in India, after quarrelling with her family, impulsively swallowed 50 mL of phorate. Immediately she lost consciousness and was taken to hospital arriving about nine hours after her suicide attempt. After five days of deep coma, the patient started to recover and was fully conscious by day 15 [44].

The reason behind the elevated risk of mood disorder in farming populations is unclear. The neurotoxic effects of high-level acute poisoning are well known and involve inhibition of the enzyme AChE, provoking changes in peripheral, autonomic, and CNS function resulting in a group of physical, cognitive, and psychiatric symptoms. Nevertheless, organophosphorus pesticide (OP) disrupts many other neurotransmitters, and some of these are entailed in mood regulation, such as serotonin. This could clarify the connection between pesticide exposure and mood disorders seen in earlier studies [39,40].

Compared to other occupational groups was reported that workers exposed to OP had a high incidence of depression and anxiety [32,33], but, to the best of our knowledge, our case is the first in which phorate was ingested intentionally.

Deaths from acute OP intoxication usually result from the depression of the CNS respiratory system and respiratory failure caused by a combination of bronchoconstriction and excessive respiratory secretions [41,42].

Chronic renal failure is one of the problems manifested in clinical follow-up of the patients and causes an increase in OP intoxication mortality [43]. In this specific case, after the exposition for several years to phorate, the patient developed depressive disorders, and the postmortem histological findings showed chronic kidney damage.

A significant number of studies demonstrated that phorate induces structural DNA modification with cellular damage in cultured human and animal cells [50,51]. The production of reactive oxygen species (ROS) in the vicinity of the protein molecule and the strong binding of phorate to the proteins induce the protein damage [52]. Several studies confirmed the ROS generating capability of phorate both extracellularly and intracellularly, with mitochondrial damage. In fact, the production of ROS induced mitochondrial damage, which suggests phorate toxicity in exposed cells [53,54]. Damage to lysosomal membranes is known to release lysosome protease into intracellular spaces, affecting neighboring cells, and triggers cell death due to necrosis [55].

These structural cell alterations have been highlighted in the kidney, a target organ of phorate as highlighted both in animal and human studies. In general acute renal failure is due to the oxidative stress, where OP directly damages renal tubules and renal parenchyma, and to the myoglobinuria caused by muscle fasciculation [56–58].

Chronic exposure to phorate is experimentally studied by histopathological changes observed in the proximal renal tubules; similar changes were observed in experiments conducted on animal and human models [33,34].

Qi et al. [46] investigated the protective effect of quercetin against the combined toxic action induced by the mixture of four organophosphate pesticides (dimethoate, acephate, dichlorvos, and phorate) given to the rats for 90 days, using metabonomics detected in rat urine. The authors showed the toxicity induced by chronic exposure to low-level mixture of OPs on the metabolism of fatty acids, energy and sex hormones, antioxidant defense system, DNA damage, and kidney function.

In a very similar study, Du et al. [46] in the plasma rats carried out a metabolomic analysis of the combined toxic action of long-term low-level exposure to a mixture of four organophosphate pesticides (dichlorvos, dimethoate, acephate, and Phorate) given for 24 weeks. The results highlighted that the mixture of OP pesticides induced oxidative stress, renal dysfunction, disturbed the metabolism of lipids and amino acids, and the thyroid gland. Observational histopathological changes in the kidney are marked by renal tubular epithelial cell swelling and granular degeneration, which were observed in the low-dose group 12 weeks after treatment.

Histopathological changes in the kidney were seen in the high- and middle-dose groups than those in the low-dose group 12 weeks after treatment. The main characteristic findings were the renal tubular epithelial cell swelling, granular degeneration, and vacuolar degeneration.

The same study was carried out by Sun et al. [47], analyzing the urine of rats investigating the toxic effect of long-term and low-level exposure to phorate using a metabolomics approach ultra-performance liquid chromatography-mass spectrometry (UPLC-MS). Male Wistar rats were administered low doses of 0.05, 0.15, or 0.45 mg/kg body weight (BW) phorate daily through water for 24 weeks consecutively. The levels of creatinine (CR) and urea nitrogen (BUN) were significantly elevated in the high-dose group, indicating kidney damage after exposure to phorate.

Li et al. [48] examined the effect of quercetin against a mixture of four organophosphates (dichlorvos, acephate, dimethoate, and phorate) inducing nephrotoxicity in rats, given for 90 days. Similar to the previous studies, the histopathological examination performed on the kidney sections (H&E method) showed extensive cell vacuolar denaturation and desquamation of the epithelial lining of the tubules, renal damage by impairing the reabsorption capacity of the proximal tubules, and by decreasing glomerular filtration rate.

Mohssen [49], in his study, evaluated biochemical (serum creatinine) and histopathological kidney alterations in male Swiss albino mice, Mus musculus, caused by subchronic inhalation of the recommended field dose of Phorate (20 kg/ha). In particular, from the second to twelfth week of exposure, the kidney, in the cortical tubules, showed at the beginning widespread cloudy degeneration, with a few foci of hydropic degeneration until the blockage of the lumen by necrotic cells. In the interstitial spaces, red blood cells and macrophages were present in the first weeks, with an increase of the chronic inflammatory infiltrate associated with foci of hemorrhage in the cortex. Kidney damage is caused by the ability of phorate to induce inflammation-inducing the formation of reactive oxygen, damage to the mitochondria, damage to the cell membrane and proteins, including enzymes, which finally result in loss of membrane fluidity and function. These alterations are associated with an early injury to cellular membranes after exposure to different toxins that induce plasma membrane injury disturbances.

Moreover, phorate exhibits cellular DNA damage and activation of apoptosis-related p53, caspase 3 and 9 genes, inducing the generation of intracellular ROS that induce reduced activities of the antioxidant enzymes catalase (CAT), glutathione (GSH) and lipid peroxidation (LPO), which were observed in the kidney of rats exposed to phorate [36]. Male Wistar rats exposed to phorate at varying oral doses of 0.046, 0.092, or 0.184 mg phorate/kg BW for 14 days showed, after histological investigations of tissues, changes in kidney function such as renal blood flow, concentrating substances, and biotransformation of the parent compounds make this tissue sensitive to a variety of toxins. The greater Bowman's space showed with infiltration of renal parenchyma by inflammatory leukocytes, dilated blood vessels, and renal necrosis; similar results were shown by Mohssen [49].

5. Conclusions

Through this case and the data reported in the literature, the authors want to point out that: (a) the use of OP and phorate should be limited in time because increases symptoms of psychological distress, including suicidal thoughts; (b) the OP, particularly after the prolonged exposition, have a toxic effect on the kidney that is histologically notable; (c) in this case, the cause of death was attributed to respiratory failure with pulmonary dysfunction due to an acute cholinergic crisis; (d) phorate induces DNA structural alterations and cellular damage in cultured human cells and in animal cells determining qualitative changes in tissues in the kidney.

Moreover, since there are, to date, few reports of similar deaths, our report provides useful information regarding this particular kind of death. All aspects of the forensic death investigation triad -investigation (history), pathology, and laboratory results—are essential and must be evaluated in context with one another. In this regard, the toxicological investigation was decisive, allowing us to identify the stomach's substance and was analyzed after liquid/liquid extraction by GC–MS analysis (single quadrupole).

Since extremely hazardous neurotoxic pesticides are still ordinally used in developing countries, further research on mental health is crucial to make an important preventive action by governmental and international bodies.

Author Contributions: Conceptualization, A.M. and C.L.; methodology, M.E., and V.R.; software, F.A. and N.D.N.; validation, N.D.N., A.M., and M.S.; formal analysis, C.L., V.R., and G.C.; investigation, G.C.; resources, AM., M.S. and C.L.; data curation, V.R., and N.D.N.; writing—original draft preparation, A.M. and C.L., writing—review and editing, A.M. and C.L.; supervision, V.R., and A.M. All authors have read and agreed to the published version of the manuscript.

Funding: This research received no external funding.

Institutional Review Board Statement: All procedures performed in the study were in accordance with the ethical standards of the institution and with the 1964 Helsinki Declaration and its later amendments or comparable ethical standards.

Informed Consent Statement: Informed consent was obtained from the relatives.

Data Availability Statement: All data are included in the main text.

Acknowledgments: The authors thank the Scientific Bureau of the University of Catania for language support.

Conflicts of Interest: The authors declare no conflict of interest.

References

1. World Health Organization. *Pesticide Residues in Food-2004: Evaluations 2004: Part II-Toxicological*; WHO: Geneva, Switzerland, 2006; Volume 178.
2. Snider, T.H.; McGarry, K.G.; Babin, M.C.; Jett, D.A.; Platoff, G.E., Jr.; Yeung, D.T. Acute toxicity of Phorate oxon by oral gavage in the Sprague-Dawley rat. *Fundam. Toxicol. Sci.* **2016**, *3*, 195–204. [CrossRef] [PubMed]
3. Khatiwada, S.; Tripathi, M.; Pokharel, K.; Acharya, R.; Subedi, A. Ambiguous Phorate granules for sesame seeds linked to accidental organophosphate fatal poisoning. *JNMA J. Nepal Med. Assoc.* **2012**, *52*, 49–51. [CrossRef] [PubMed]
4. Meltzer, H.; Griffiths, C.; Brock, A.; Rooney, C.; Jenkins, R. Patterns of suicide by occupation in England and Wales: 2001–2005. *Br. J. Psychiatry* **2008**, *193*, 73–76. [CrossRef] [PubMed]
5. Miller, K.; Burns, C. Suicides on farms in South Australia, 1997–2001. *Aust. J. Rural Health* **2008**, *16*, 327–331. [CrossRef]
6. Lee, W.J.; Alavanja, M.C.; Hoppin, J.A.; Rusiecki, J.A.; Kamel, F.; Blair, A.; Sandler, D.P. Mortality among pesticide applicators exposed to chlorpyrifos in the Agricultural Health Study. *Environ. Health Perspect.* **2007**, *115*, 528–534. [CrossRef]
7. Gregoire, A. The mental health of farmers. *Occup. Med.* **2002**, *52*, 471–476. [CrossRef]
8. MacFarlane, E.; Benke, G.; Del Monaco, A.; Sim, M.R. Cancer incidence and mortality in a historical cohort of Australian pest control workers. *Occup. Environ. Med.* **2009**, *66*, 818–823. [CrossRef] [PubMed]
9. MacFarlane, E.; Benke, G.; Del Monaco, A.; Sim, M.R. Causes of death and incidence of cancer in a cohort of Australian pesticide-exposed workers. *Ann. Epidemiol.* **2010**, *20*, 273–280. [CrossRef] [PubMed]
10. Torchio, P.; Lepore, A.R.; Corrao, G.; Comba, P.; Settimi, L.; Belli, S.; Magnani, C.; di Orio, F. Mortality study on a cohort of Italian licensed pesticide users. *Sci. Total Environ.* **1994**, *149*, 183–191. [CrossRef]
11. Parrón, T.; Hernández, A.F.; Villanueva, E. Increased risk of suicide with exposure to pesticides in an intensive agricultural area. A 12-year retrospective study. *Forensic Sci. Int.* **1996**, *79*, 53–63. [CrossRef]
12. Stallones, L. Suicide and potential occupational exposure to pesticides, Colorado, 1990–1999. *J. Agromed.* **2006**, *11*, 107–112. [CrossRef] [PubMed]
13. London, L.; Flisher, A.J.; Wesseling, C.; Mergler, D.; Kromhout, H. Suicide and exposure to organophosphate insecticides: Cause or effect? *Am. J. Ind. Med.* **2005**, *47*, 308–321. [CrossRef] [PubMed]

14. Alavanja, M.C.; Sandler, D.P.; McMaster, S.B.; Zahm, S.H.; McDonnell, C.J.; Lynch, C.F.; Blair, F.A. The Agricultural Health Study. *Environ. Health Perspect.* **1996**, *104*, 362–369. [CrossRef] [PubMed]
15. Purdue, M.P.; Hoppin, J.A.; Blair, A.; Dosemeci, M.; Alavanja, M.C. Occupational exposure to organochlorine insecticides and cancer incidence in the Agricultural Health Study. *Int. J. Cancer* **2007**, *120*, 642–649. [CrossRef]
16. Hou, L.; Lee, W.J.; Rusiecki, J.; Hoppin, J.A.; Blair, A.; Bonner, M.R.; Lubin, J.H.; Samanic, C.; Sandler, D.P.; Dosemeci, M.; et al. Pendimethalin exposure and cancer incidence among pesticide applicators. *Epidemiology* **2006**, *17*, 302. [CrossRef]
17. Andreotti, G.; Freeman, L.E.B.; Hou, L.; Coble, J.; Rusiecki, J.; Hoppin, J.A.; Silverman, D.T.; Alavanja, M.C. Agricultural pesticide use and pancreatic cancer risk in the Agricultural Health Study cohort. *Int. J. Cancer* **2009**, *124*, 2495–2500. [CrossRef]
18. Lee, W.J.; Sandler, D.P.; Blair, A.; Samanic, C.; Cross, A.J.; Alavanja, M.C.R. Pesticide use and colorectal cancer risk in the Agricultural Health Study. *Int. J. Cancer* **2007**, *121*, 339–346. [CrossRef]
19. van Bemmel, D.M.; Visvanathan, K.; Beane Freeman, L.E.; Coble, J.; Hoppin, J.A.; Alavanja, M.C. S-ethyl-N, N-dipropylthiocarbamate exposure and cancer incidence among male pesticide applicators in the Agricultural Health Study: A prospective cohort. *Environ. Health Perspect.* **2008**, *116*, 1541–1546. [CrossRef]
20. Rusiecki, J.A.; Patel, R.; Koutros, S.; Beane-Freeman, L.; Landgren, O.; Bonner, M.R.; Alavanja, M.C. Cancer incidence among pesticide applicators exposed to permethrin in the Agricultural Health Study. *Environ. Health Perspect.* **2009**, *117*, 581–586. [CrossRef]
21. Bonner, M.R.; Coble, J.; Blair, A.; Beane Freeman, L.E.; Hoppin, J.A.; Sandler, D.P.; Alavanja, M.C. Malathion exposure and the incidence of cancer in the agricultural health study. *Am. J. Epidemiol.* **2007**, *166*, 1023–1034. [CrossRef]
22. Kang, D.; Park, S.K.; Beane-Freeman, L.; Lynch, C.F.; Knott, C.E.; Sandler, D.P.; Hoppin, J.A.; Dosemeci, M.; Coble, J.; Lubin, J.; et al. Cancer incidence among pesticide applicators exposed to trifluralin in the Agricultural Health Study. *Environ. Res.* **2008**, *107*, 271–276. [CrossRef] [PubMed]
23. Rajkumar, S.V. MGUS and smoldering multiple myeloma: Update on pathogenesis, natural history, and management. *Hematology* **2005**, *2005*, 340–345. [CrossRef] [PubMed]
24. Koutros, S.; Lynch, C.F.; Ma, X.; Lee, W.J.; Hoppin, J.A.; Christensen, C.H.; Sandler, D.P. Heterocyclic aromatic amine pesticide use and human cancer risk: Results from the US Agricultural Health Study. *Int. J. Cancer* **2009**, *124*, 1206–1212. [CrossRef] [PubMed]
25. Lee, W.J.; Blair, A.; Hoppin, J.A.; Lubin, J.H.; Rusiecki, J.A.; Sandler, D.P.; Dosemeci, M.; Alavanja, M.C.R. Cancer incidence among pesticide applicators exposed to chlorpyrifos in the Agricultural Health Study. *J. Natl. Cancer Inst.* **2004**, *96*, 1781–1791. [CrossRef] [PubMed]
26. Mahajan, R.; Blair, A.; Coble, J.; Lynch, C.F.; Hoppin, J.A.; Sandler, D.P.; Alavanja, M.C. Carbaryl exposure and incident cancer in the Agricultural Health Study. *Int. J. Cancer* **2007**, *121*, 1799–1805. [CrossRef]
27. Kashyap, S.K.; Jani, J.P.; Saiyed, H.N.; Gupta, S.K. Clinical effects and cholinesterase activity changes in workers exposed to Phorate (Thimet). *J. Environ. Sci. Health B* **1984**, *19*, 479–489. [CrossRef]
28. Rapisarda, V.; Ledda, C.; Matera, S.; Fago, L.; Arrabito, G.; Falzone, L.; Marconi, A.; Libra, M.; Loreto, C. Absence of t (14; 18) chromosome translocation in agricultural workers after short-term exposure to pesticides. *Mol. Med. Rep.* **2017**, *15*, 3379–3382. [CrossRef]
29. Casida, J.E.; Quistad, G.B. Organophosphate toxicology: Safety aspects of nonacetylcholinesterase secondary targets. *Chem. Res. Toxicol.* **2004**, *17*, 983–998. [CrossRef]
30. Vinceti, M.; Filippini, T.; Violi, F.; Rothman, K.J.; Costanzini, S.; Malagoli, C.; Wise, L.A.; Odone, A.; Signorelli, C.; Iacuzio, L.; et al. Pesticide exposure assessed through agricultural crop proximity and risk of amyotrophic lateral sclerosis. *Environ. Health* **2017**, *16*, 91. [CrossRef]
31. Costa, C.; Rapisarda, V.; Catania, S.; Di Nola, C.; Ledda, C.; Fenga, C. Cytokine patterns in greenhouse workers occupationally exposed to α-cypermethrin: An observational study. *Environ. Toxicol. Pharmacol.* **2013**, *36*, 796–800. [CrossRef]
32. Farahat, T.M.; Abdelrasoul, G.M.; Amr, M.M.; Shebl, M.M.; Farahat, F.M.; Anger, W.K. Neurobehavioural effects among workers occupationally exposed to organophosphorous pesticides. *Occup. Environ. Med.* **2003**, *60*, 279–286. [CrossRef] [PubMed]
33. Kamel, F.; Hoppin, J.A. Association of pesticide exposure with neurologic dysfunction and disease. *Environ. Health Perspect.* **2004**, *112*, 950–958. [CrossRef] [PubMed]
34. Martínez-Valenzuela, C.; Waliszewski, S.M.; Amador-Muñoz, O.; Meza, E.; Calderón-Segura, M.E.; Zenteno, E.; Huichapan-Martínez, J.; Caba, M.; Félix-Gastélum, R.; Longoria-Espinoza, R. Aerial pesticide application causes DNA damage in pilots from Sinaloa, Mexico. *Environ. Sci. Pollut. Res. Int.* **2017**, *24*, 2412–2420. [CrossRef] [PubMed]
35. Numan, I.T.; Hassan, M.Q.; Stohs, S.J. Endrin-induced depletion of glutathione and inhibition of glutathione peroxidase activity in rats. *Gen. Pharmacol. Vasc. Syst.* **1990**, *21*, 625–628. [CrossRef]
36. Bakheet, S.A.; Attia, S.M.; Al-Rasheed, N.M.; Al-Harbi, M.M.; Ashour, A.E.; Korashy, H.M.; Abd-Allah, A.R.; Saquib, Q.; Al-Khedhairy, A.A.; Musarrat, J. Salubrious effects of dexrazoxane against teniposide-induced DNA damage and programmed cell death in murine marrow cells. *Mutagenesis* **2011**, *26*, 533–543. [CrossRef]
37. Mohanty, G.; Mohanty, J.; Nayak, A.K.; Mohanty, S.; Dutta, S.K. Application of comet assay in the study of DNA damage and recovery in rohu (*Labeo rohita*) fingerlings after an exposure to Phorate, an organophosphate pesticide. *Ecotoxicology* **2011**, *20*, 283–292. [CrossRef]
38. Moher, D.; Shamseer, L.; Clarke, M.; Ghersi, D.; Liberati, A.; Petticrew, M.; Shekelle, P.; Stewart, L.A. Preferred reporting items for systematic review and meta-analysis protocols (PRISMA-P) 2015 statement. *Syst. Rev.* **2015**, *4*, 1. [CrossRef]

39. Moyer, R.A.; McGarry Jr, K.G.; Babin, M.C.; Platoff Jr, G.E.; Jett, D.A.; Yeung, D.T. Kinetic analysis of oxime-assisted reactivation of human, Guinea pig, and rat acetylcholinesterase inhibited by the organophosphorus pesticide metabolite Phorate oxon (PHO). *Pestic. Biochem. Physiol.* **2018**, *145*, 93–99. [CrossRef]
40. Busardo, F.P.; Pichini, S.; Pellegrini, M.; Montana, A.; Lo Faro, A.F.; Zaami, S.; Graziano, S. Correlation between blood and oral fluid psychoactive drug concentrations and cognitive impairment in driving under the influence of drugs. *Curr. Neuropharmacol.* **2018**, *16*, 84–96. [CrossRef]
41. Zafar, R.; Munawar, K.; Nasrullah, A.; Haq, S.; Ghazanfar, H.; Sheikh, A.B.; Khan, A.Y. Acute Renal Failure due to Organophosphate Poisoning: A Case Report. *Cureus* **2017**, *9*. [CrossRef]
42. Barbera, N.; Montana, A.; Indorato, F.; Arbouche, N.; Romano, G. Evaluation of the role of toxicological data in discriminating between H2S femoral blood concentration secondary to lethal poisoning and endogenous H2S putrefactive production. *J. Forensic Sci.* **2017**, *62*, 392–394. [CrossRef] [PubMed]
43. Maślińska, D.; Lewandowska, I.; Prokopczyk, J. Effect of Prolonged Acetylcholinesterase Inhibition on Postnatal Brain Development in Rabbit I. Level of Serotonin in Different Brain Regions. In *Experimental and Clinical Neuropathology*; Springer: Berlin/Heidelberg, Germany, 1981; pp. 52–55.
44. Peter, J.V.; Prabhakar, A.T.; Pichamuthu, K. In-laws, insecticide—And a mimic of brain death. *Lancet* **2008**, *371*, 622. [CrossRef]
45. Qi, L.; Cao, C.; Hu, L.; Chen, S.; Zhao, X.; Sun, C. Metabonomic analysis of the protective effect of quercetin on the toxicity induced by mixture of organophosphate pesticides in rat urine. *Hum. Exp. Toxicol.* **2017**, *36*, 494–507. [CrossRef] [PubMed]
46. Du, L.; Li, S.; Qi, L.; Hou, Y.; Zeng, Y.; Xu, W.; Wang, H.; Zhao, X.; Sun, C. Metabonomic analysis of the joint toxic action of long-term low-level exposure to a mixture of four organophosphate pesticides in rat plasma. *Mol. Biosyst.* **2014**, *10*, 1153–1161. [CrossRef] [PubMed]
47. Sun, X.; Xu, W.; Zeng, Y.; Hou, Y.; Guo, L.; Zhao, X.; Sun, C. Metabonomics evaluation of urine from rats administered with Phorate under long-term and low-level exposure by ultra-performance liquid chromatography-mass spectrometry. *J. Appl. Toxicol.* **2014**, *34*, 176–183. [CrossRef]
48. Li, S.; Can Cao, H.; Shuang, Y.; Lei, Q.; Xiujuan, Z.; Changhao, S. Effect of quercetin against mixture of four organophosphate pesticides induced nephrotoxicity in rats. *Xenobiotica* **2016**, *46*, 225–233. [CrossRef]
49. Mohssen, M. Biochemical and histopathological changes in serum creatinine and kidney induced by inhalation of Thimet (Phorate) in male Swiss albino mouse, Mus musculus. *Environ. Res.* **2001**, *87*, 31–36. [CrossRef]
50. Saquib, Q.; Musarrat, J.; Siddiqui, M.A.; Dutta, S.; Dasgupta, S.; Giesy, J.P.; Al-Khedhairy, A.A. Cytotoxic and necrotic responses in human amniotic epithelial (WISH) cells exposed to organophosphate insecticide Phorate. *Mutat. Res. Toxicol. Environ. Mutagen.* **2012**, *744*, 125–134. [CrossRef]
51. Saquib, Q.; Al-Khedhairy, A.A.; Siddiqui, M.A.; Roy, A.S.; Dasgupta, S.; Musarrat, J. Preferential binding of insecticide Phorate with sub-domain IIA of human serum albumin induces protein damage and its toxicological significance. *Food Chem. Toxicol.* **2011**, *49*, 1787–1795. [CrossRef]
52. Abdollahi, M.; Ranjbar, A.; Shadnia, S.; Nikfar, S.; Rezaiee, A. Pesticides and oxidative stress: A review. *Med. Sci. Monit.* **2004**, *10*, 141–147.
53. Fadok, V.A.; Voelker, D.R.; Campbell, P.A.; Cohen, J.J.; Bratton, D.L.; Henson, P.M. Exposure of phosphatidylserine on the surface of apoptotic lymphocytes triggers specific recognition and removal by macrophages. *J. Immunol.* **1992**, *148*, 2207–2216. [PubMed]
54. Martin, S.; Reutelingsperger, C.P.; McGahon, A.J.; Rader, J.A.; Van Schie, R.C.; LaFace, D.M.; Green, D.R. Early redistribution of plasma membrane phosphatidylserine is a general feature of apoptosis regardless of the initiating stimulus: Inhibition by overexpression of Bcl-2 and Abl. *J. Exp. Med.* **1995**, *182*, 1545–1556. [CrossRef] [PubMed]
55. Leist, M.; Jäättelä, M. Four deaths and a funeral: From caspases to alternative mechanisms. *Nat. Rev. Mol. Cell Biol.* **2001**, *2*, 589–598. [CrossRef] [PubMed]
56. Karalliedde, L.; Senanayake, N. Organophosphorus insecticide poisoning. *Br. J. Anaesth.* **1989**, *63*, 736–750. [CrossRef] [PubMed]
57. Betrosian, A.; Balla, M.; Kafiri, G.; Kofinas, G.; Makri, R.; Kakouri, A. Multiple systems organ failure from organophosphate poisoning. *J. Toxicol. Clin. Toxicol.* **1995**, *33*, 257–260. [CrossRef]
58. Saquib, Q.; Sabry, M.; Attia, M.A.; Siddiqui, M.; Aboul, S.; Abdulaziz, A.; Al, K.; John, P.G.; Javed, M. Phorate-induced oxidative stress, DNA damage and transcriptional activation of p53 and caspase genes in male Wistar rats. *Toxicol. Appl. Pharmacol.* **2012**, *259*, 54–65. [CrossRef]

 *healthcare* 

Article

A Descriptive Study on Causes of Death in Hospitalized Patients in an Acute General Hospital of Southern Italy during the Lockdown due to Covid-19 Outbreak

Pasquale Mascolo [1], Alessandro Feola [1,*], Carmen Sementa [2], Sebastiano Leone [3], Pierluca Zangani [1], Bruno Della Pietra [1] and Carlo Pietro Campobasso [1]

[1] Department of Experimental Medicine, University of Campania "Luigi Vanvitelli", Via Luciano Armanni 5, 80138 Naples, Italy; mascolopasquale2@gmail.com (P.M.); pierluca.zangani@unicampania.it (P.Z.); bruno.dellapietra@unicampania.it (B.D.P.); carlopietro.campobasso@unicampania.it (C.P.C.)

[2] Unit of Legal Medicine, AORN "San Giuseppe Moscati", Contrada Amoretta, 83100 Avellino, Italy; medicinalegale@aosgmoscati.av.it

[3] Unit of Infectious Disease, AORN "San Giuseppe Moscati", Contrada Amoretta, 83100 Avellino, Italy; sebastianoleone@yahoo.it

* Correspondence: alessandro.feola@unicampania.it; Tel.: +39-081-5666018

Abstract: (1) Background: All deaths that occurred in a hospital of Southern Italy ("San Giuseppe Moscati" Hospital of Avellino) with medium jurisdiction (up to 425,000 citizens approximately) in the period from 9 March to 4 May 2020 were analyzed. The primary endpoint of the study was to analyze the causes of death in the period study. Secondary endpoints included: (1) the assessment of overall mortality in the emergency period compared with the same period of the past years (2018–2019) in the jurisdiction area; (2) the assessment of the amounts of deaths with positive and negative reverse transcription-polymerase chain reaction (RT-PCR) of nasopharyngeal and oropharyngeal swabs; (3) the frequency of clinical and radiological features consistent with Covid-19 infection in negative RT-PCR cases. (2) Methods: Patients' information and laboratory data were collected through the computerized medical record system (My Hospital, Italy) used for the clinical management of all referring patients. Epidemiological, clinical, and radiological data were reviewed along with the results of nasopharyngeal and oropharyngeal swabs. (3) Results: From 9 March to 4 May 2020, 140 deaths (87 males, 53 females) from all causes occurred in total at "San Giuseppe Moscati" Hospital, of which 32 deaths were Covid-19 related. (4) Conclusions: The excess of mortality could be higher than the one reported in the official epidemiological surveys. False negative cases can have a distorting effect on the assessment of the real mortality rate and the excess mortality. Furthermore, many who died from Covid-19 were likely never tested or they had false negative RT-PCR results. Other victims probably died from causes indirectly related to Covid-19.

Keywords: COVID-19; mortality; RT-PCR; chest computed tomography

Citation: Mascolo, P.; Feola, A.; Sementa, C.; Leone, S.; Zangani, P.; Della Pietra, B.; Campobasso, C.P. A Descriptive Study on Causes of Death in Hospitalized Patients in an Acute General Hospital of Southern Italy during the Lockdown due to Covid-19 Outbreak. *Healthcare* **2021**, *9*, 119. https://doi.org/10.3390/healthcare9020119

Received: 22 November 2020
Accepted: 20 January 2021
Published: 25 January 2021

1. Background

COVID-19 (coronavirus disease) is still a global health emergency [1,2]. It is a pneumonia caused by a novel enveloped RNA betacoronavirus [3] that has currently been named severe acute respiratory syndrome coronavirus 2 (SARS-CoV-2) [4]. This infection is characterized most frequently by fever, dry cough, and dyspnea [5–8]. The symptoms of Covid-19 infection commonly appear after an incubation period of approximately 5.2 days [9]. However, the asymptomatic transmission is considered the Achilles' heel of current strategies to control the infection [10].

Actually, it is still extremely difficult to estimate the total number of infected people, because asymptomatic patients with very mild symptoms might not be tested and, therefore, not identified [11]. In fact, the gold standard for the diagnosis of Covid-19 is by reverse transcription-polymerase chain reaction (RT-PCR) [12], but delay in the results can occur because of variability in the swabs processing [13]. The total positive rate of RT-PCR for throat swab samples amounts to about 30% to 60% at initial evaluation due to the limitations of sample collections and transportation and kit performance [12,14]. Therefore, negative nasopharyngeal and oropharyngeal swab does not rule out Covid-19 infection [15].

The chest computed tomography (CT) can have a relevant role in cases with negative swabs. Chest CT scan shows peculiar features in almost all Covid-19 patients, including ground-glass opacities (GGOs), multifocal patchy consolidation, and/or interstitial changes with a peripheral distribution [16]. Chest radiological pattern was found to be useful in confirming the Covid-19 diagnosis [17,18]. However, atypical or negative chest CT findings do not exclude a Covid-19 infection, especially in the first three days, just because chest imaging may also lead to both false-negative and false-positive results [19].

The purpose of this research study is to analyze all deaths that occurred in a hospital of Southern Italy ("San Giuseppe Moscati" Hospital of Avellino) with medium jurisdiction (up to 425,000 citizens, approximately) in the period from 9 March 2020 to 4 May 2020. This is the period of pandemic lockdown in Italy during which the government imposed a national quarantine, restricting the movement of the population except for necessity, work, and health circumstances.

Avellino is located in the Campania region that was less affected by Covid-19 compared with Lombardy region in Northern Italy [20]. However, the real amount of deaths Covid-19 related could be even higher due to doubts raised about the magnitude of the infectious disease on our small community. Although, at the beginning of May, the Campania region comprised 1.25% of all Covid-19 related deaths (336 out of 29,079 in total), the Italian National Institute of Statistics (ISTAT) reports an excess mortality from coronavirus pandemic of 26,350 deaths in March and 16,283 in April all over the country [21]. These deaths are those above the average number of deaths that occurred in the previous years from 2015 to 2019 in Italy. The excess mortality was mostly due to males in the age range of 60–69 years (95%) followed soon after by those in the age range of 70–79 years (80%) and in the age range of 80–89 (57%). According to the ISTAT report [21], in Italy, the excess mortality from Covid-19 was approximately of 49.4% in March, 36.6% in April, and 3.9% in May for territories with high risk of infection. The excess mortality of Southern Italy was of 5.5% in March and 4.0% in April, which is relatively low if compared with the excess mortality of Northern Italy (96.4% in March, 71.7% in April and 3.2% in May).

Secondary endpoints of the study are: (1) the assessment of overall mortality in the emergency period compared with the same period of the past years (2018–2019) in the jurisdiction area; (2) the assessment of the amounts of deaths with positive and negative RT-PCR of nasopharyngeal and oropharyngeal swabs; (3) the frequency of clinical and radiological features consistent with Covid-19 infection in negative RT-PCR cases.

2. Materials and Methods

2.1. Subjects and Subgroup Formation

From 9 March 2020 to 4 May 2020, 140 deaths (87 males, 53 females) from all causes occurred in total at "San Giuseppe Moscati" Hospital, a 572-bed hospital located in Avellino (Italy). Patients' information and laboratory data were collected through the computerized medical record system (My Hospital, Italy) used for the clinical management of all referring patients. Epidemiological, clinical, and radiological data were reviewed along with the results of nasopharyngeal and oropharyngeal swabs. The number of deaths at the "San Giuseppe Moscati" Hospital has been compared with the overall mortality occurred in the past years 2018–2019. No nosocomial outbreak of infection in the hospital was recorded at the same time. Data collection and analysis were approved by the Medical Ethics Committee of the "San Giuseppe Moscati" Hospital in accordance with international and institutional ethics guidelines as well as Italian legislation dealing with data protection.

2.2. Radiological Investigation

Image analysis was performed by one radiologist who was not aware of the RT-PCR results. The epidemiological history and the clinical symptoms (fever and/or dry cough) were the only data available for the radiologist. Radiological patterns of chest CT scan were then classified as typical or atypical for Covid-19 according to international radiological guidelines [13,22].

GGOs, particularly on peripheral and lower lobes, multifocal patchy consolidation, and crazy paving, were considered typical Covid-19 radiological patterns [18]. Atypical CT findings have included pleural or pericardial effusion, cavitation, and lymphadenopathies [23].

All CT images were obtained with CT systems (Siemens SOMATOM Sensation 16) with patients in supine position. The main scanning parameters were as follows: tube voltage, 100 kVp; automatic tube current modulation; tube current, 30–70 mAs; pitch, 0.99–1.22 mm; matrix, 5,123,512; slice thickness, 3 mm; and field of view, 350 mm 3350 mm. All images were then reconstructed with a (16 slices) slice thickness of 0.625–1.250 mm with the same increment.

2.3. Molecular Test

Among the 140 deaths, 108 nasopharyngeal and oropharyngeal swabs were carried out in total. Most of the swabs were taken from patients admitted to the hospital, except in 9 cases, where the throat samples were collected soon after death and, in a single case, 24 h after death. All the post mortem swabs gave negative RT-PCR results.

The RT-PCR assays were performed by using TaqMan One-Step RT-PCR Kits (Shanghai Huirui Biotechnology (Shanghai, China) or Shanghai BioGerm Medical Biotechnology (Shanghai, China)), which have been approved for use by the China Food and Drug Administration. The sensitivity of the RT-PCR diagnostic test has been reported to be 0.777 (95% CI: 0.715, 0.849), while the specificity was 0.988 (95% CI: 0.933, 1.000) with a conspicuous rate of false negative results, likely missing between 15% and 29% of patients with Covid-19 [24].

Among the 108 throat swabs, 76 cases had negative RT-PCR for Covid-19 (study group #1) and, in 32 deaths, RT-PCR were positive (study group #2). These 32 deaths were considered to be cases suffering from Covid-19 because laboratory confirmation occurred according to WHO criteria for the diagnosis of Covid-19 [2]. For the additional 32 deaths out of 140 total, no swabs were performed, as the cause of death was something other than Covid-19. These 32 deaths (study group #3) were excluded from the present survey, as the cause of death was certainly something other than Covid-19. They were mostly natural deaths due to diseases of the cardiovascular system (myocardial infarction, stroke, etc.) and the nervous system (brain cancer, muscular dystrophy, etc.) and some traumatic deaths due to traffic injuries, accidental falls, poisoning, etc.

Figure 1 reports the flow diagram depicting the study groups. Among the 76 cases with negative RT-PCR results (sample study #1), 28 deaths (cohort #1.1) showed chest CT features consistent with Covid-19 disease. In this cohort #1.1, the 10 post mortem swabs are included. For 48 patients out of 76 (cohort #1.2), no chest CT was available.

2.4. Statistical Analysis

A statistical analysis was performed using the $\chi 2$ test to compare the typical CT findings and the RT-PCR results in the study group. A p-value < 0.05 was considered statistically significant.

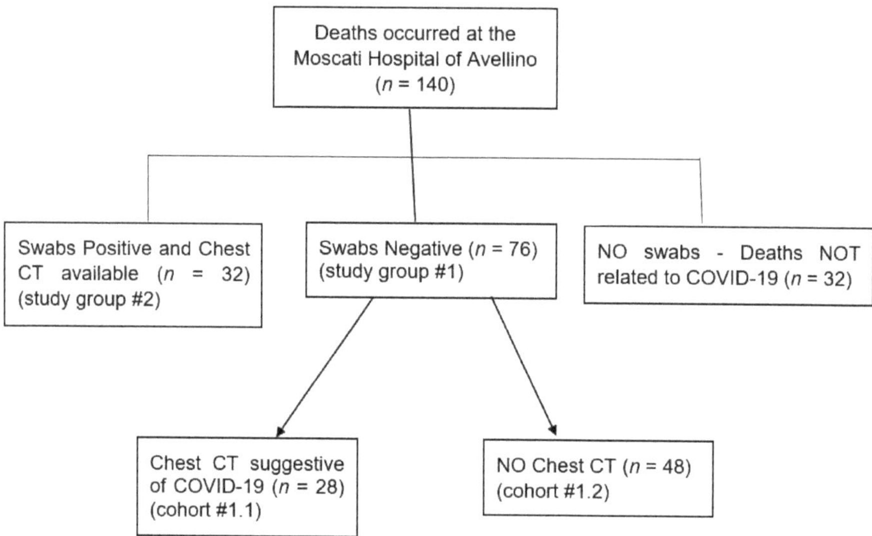

Figure 1. The flow diagram depicting the study groups.

3. Results

In the national lockdown period, 32 deaths from coronavirus out of 53 occurred at the "San Giuseppe Moscati" Hospital in Avellino (Italy).

In Figure 2, it is shown the excess mortality from Covid-19 in comparison with the number of deaths reported in 2018 and 2019. According to the hospital register, the total number of deaths reported in March and April 2020 was 66 and 68, respectively, against an average of 34.5 cases that occurred at the same period in the previous years of 2018 and 2019. Therefore, the excess mortality reported at the "San Giuseppe Moscati" Hospital of Avellino was higher than the one observed in Campania but lower than that of the territories in the Northern Italy at high risk of infection and mortality from coronavirus [21].

The present retrospective survey was performed on 140 fatal cases that occurred in the "San Giuseppe Moscati" Hospital of Avellino from 9 March to 4 May 2020.

According to the International Classification of Disease (ICD-10), causes of death involving the respiratory system were the most represented (61 cases out of 140) followed soon after by those affecting the circulatory system (33 cases) and the nervous system (13 cases). The distribution of causes of death is shown in Figure 3.

Victims were mostly males (87 out of 140 in total—62%) with an age range between 44 and 97 years old (mean age of 75 years, median age of 78 years). The cause of death in patients with positive and negative RT-PCR results but associated with CT scan features suggestive of Covid-19 (32 + 28 cases) was assessed as pneumonia in 60 cases out of 140 totally.

All 32 patients with positive RT-PCR (group #2) showed clinical and/or radiological patterns consistent with Covid-19 disease and, therefore, these cases were confirmed as Sars-CoV-2 related deaths. The main chest CT findings were represented by GGOs alone in 18 cases out of 32 (56%), multifocal patchy consolidations associated with GGOs in 4 deaths (12%), and crazy paving associated with GGOs in 10 deaths (32%). Therefore, the GGOs were found in all 32 victims (group #2). Most of these victims were males (25 out of 32) with an age range between 55 and 94 (mean age of 75 years, median age of 73 years). All the 32 victims (group #2) had comorbidity mostly represented by hypertension in 17 cases out of 32 (53%) followed by cardiovascular diseases (12 patients out of 32—38%), chronic obstructive pulmonary disease (COPD) in 9 cases (28%), and diabetes in 6 cases out of 32 (19%).

Study group #1 (76 deaths with RT-PCR negative to Covid-19) was represented by 45 males and 31 females, with a mean age of 76 years (age distribution shown in Figure 4).

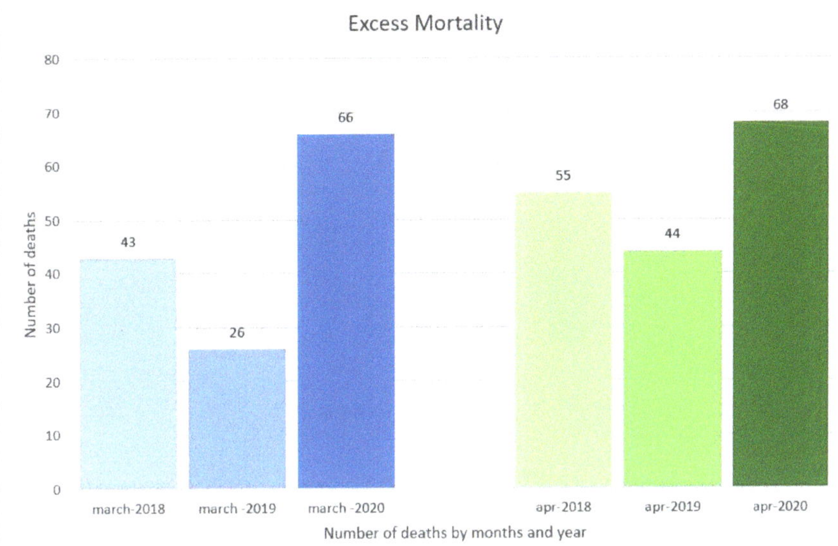

Figure 2. The excess mortality from Covid-19 in March and April 2020 in comparison with the number of deaths reported in the same period in 2018 and 2019 at the "San Giuseppe Moscati" Hospital in Avellino (Italy).

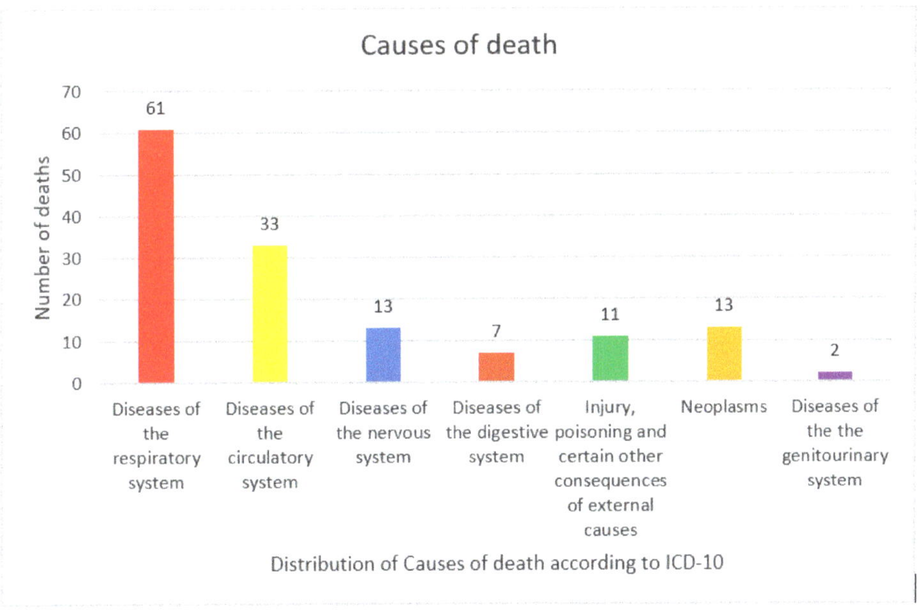

Figure 3. The distribution of causes of death among the 140 hospitalized victims.

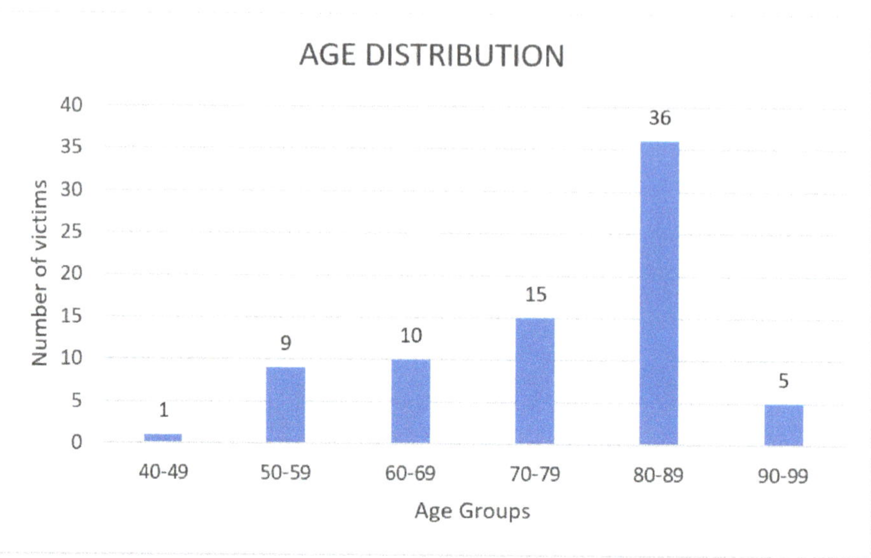

Figure 4. Age distribution of the study group #1.

Among these 76 negative cases, 28 deaths (cohort #1.1) presented clinical (fever, cough or dyspnea) and radiological findings suggestive of Covid-19 (ground-glass, consolidation, crazy paving), while in the remaining 48 deaths (cohort #1.2), no chest CT was available. For all the 76 victims, the cause of death was assessed as something other than Covid-19, although in most of the victims, severe diseases of respiratory and circulatory systems occurred (58 out of 76). However, based on the negative results of the swabs, these diseases were considered non-COVID-19 related.

The causes of deaths of 48 patients (cohort #1.2) with negative RT-PCR and no CT-chest available nor clinical symptoms Covid-19 related were mainly due to diseases of circulatory system (20 out of 48), respiratory system (10 out of 48), digestive system (2 out of 48), neoplasms (7 out of 48), nervous system (1 out 48), infectious disease other than coronavirus (3 out of 48), blunt trauma and other injuries (3 out of 48), and diseases of genitourinary system (2 out of 48).

The retrospective survey of the 28 cases belonging to cohort #1.1 (deaths with negative RT-PCR but with severe respiratory symptoms) raised initial doubts about the relationship with Sars-CoV-2 mainly because of the typical imaging features of Covid-19 pneumonia observed in chest CT. Therefore, a special focus has been performed on this cohort of 28 deaths (15 males, 13 females; median age 79 years, and age range between 65 and 87 years) with negative RT-PCR tests but with clinical and radiological imaging consistent with Covid-19 infection. According to recommendations developed for cases with negative RT-PCR but also suggestive clinical and radiological findings [13], second swabs were taken in 18 patients out of 28 in total. They were collected mostly ante-mortem but also post-mortem in 10 cases. All the RT-PCR results were again negative.

The main chest CT findings were represented by GGOs alone in 3 cases out of 28 (11%), multifocal patchy consolidations associated with GGOs in 9 deaths (32%), and crazy paving associated with GGOs in 16 deaths (57%). Therefore, GGOs were again the most common radiological abnormalities found in all the 28 victims with negative RT-PCR (cohort #1.1), although in 10 cases out of 28 in total (36%), they were associated with non-typical chest CT findings such as pleural or pericardial effusion, cavitation, and lymphoadenopathies. Statistically significant difference ($p = 0.001$, $\chi^2 = 13.82$, df = 2) was found in the comparison between the RT-PCR results and typical findings (Table 1). Therefore, 18 cases out of 28 in total (64%) showing typical findings suggestive of Covid-19 could be potentially classified as deaths from Sars-CoV2.

Table 1. Comparison between the reverse transcription-polymerase chain reaction (RT-PCR) results and typical CT findings. GGO: ground-glass opacities.

RT-PCR Results	GGOs	GGOs + Crazy Paving	GGOs + Multifocal Patchy Consolidations	p-Value
Positive RT-PCR, n. 32 (53.3%)	18 (56%)	10 (32%)	4 (12%)	0.001
Negative RT-PCR, n. 28 (46.7%)	3 (11%)	16 (57%)	9 (32%)	
Total n. 60 (100%)	21 (35%)	26 (43.3%)	13 (21.7%)	

Regarding the clinical findings that occurred among these 28 cases (cohort #1.1) with negative RT-PCR, the most common symptom was dyspnea (64%). The average duration of hospitalization was 8 days, and five patients (18%) were admitted to the intensive care unit (ICU) and underwent invasive mechanical ventilation. All the victims had severe comorbidity mostly represented by cardiovascular diseases in 15 out of 28 (54%), followed by COPD in 10 cases out of 28 (35%) and hypertension and diabetes both in 9 cases out of 28 (32%). The distribution of comorbidity among the 32 patients with positive RT-PCR (study group #2) and the 28 victims with negative RT-PCR (cohort #1.1) is summarized in Figure 5.

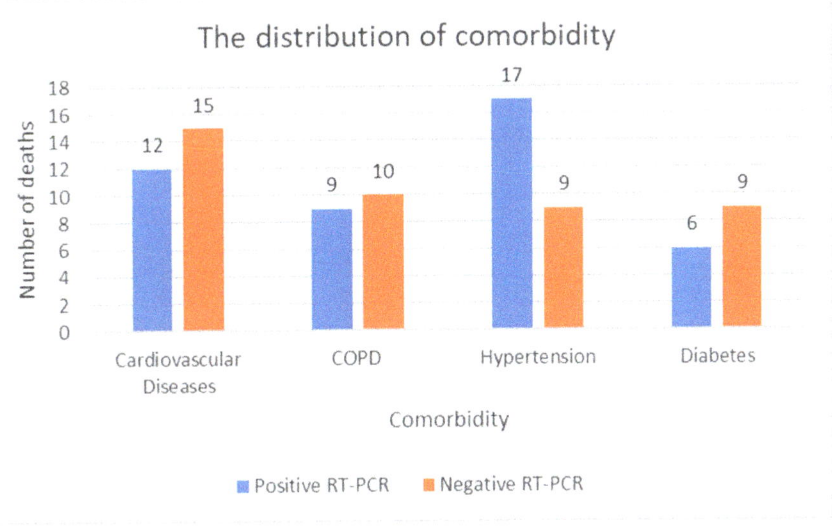

Figure 5. The distribution of comorbidity among the 32 victims with positive RT-PCR (study group #2) and the 28 deaths with negative RT-PCR (cohort #1.1).

4. Discussion

Case-fatality statistics in Italy are based on defining Covid-19 related deaths as those that occurred in patients who test positive for SARS-Cov-2 via RT-PCR, independently from preexisting diseases that may have caused death [25]. In our sample study, 32 deaths (study group #1) out of 140 in total were assessed due to Covid-19 based on positive RT-PCR and suggestive clinical and/or radiological findings. No causes of deaths other than Covid-19 occurred in this cohort of 32 Covid-19 deceased patients. The cause of death for 76 patients (study group #2) with negative RT-PCR was assessed as something other than Covid-19 and for 28 victims (cohort #1.1) who presented clinical and radiological findings consistent with Covid-19 infection. They were considered deaths not Covid-19 related just because they had negative RT-PCR results, even after a second swab taken in 18 patients out of 28 in total. The results of this retrospective survey suggest an underestimation of the effects of the Covid-19 disease on the health of a medium community such as the Avellino's one.

In this retrospective survey, the positive rate of RT-PCR assay was 30% (32 cases out of 108 swabs in total), although previous studies performed on a large sample of patients reported a greater sensibility ranging between 30 to 70% [12,14,26].

RT-PCR sensitivity depends mostly on the type of test and assay, the quality of the throat swab, and the viral load [13]. RT-PCR testing can be influenced by sampling operations, specimen source (upper or lower respiratory tract), sampling timing (different period of disease development [14,27,28], and performance of detection kits [12]. Negative results of nasopharyngeal and oropharyngeal swabs do not rule out Covid-19 disease for sure [15]. It must be taken into account that swabs can also have insufficient stability and relatively long processing time affecting RT-PCR results [12].

Therefore, false negative results of RT-PCR cannot be ignored. This could be the case of the 28 victims with negative RT-PCR (cohort #1.1). Unfortunately, SARS-Cov-2 serology was not currently available during the reference period of the first wave of the pandemic outbreak. In fact, the serologic assays to detect antibodies against SARS-CoV-2 are of great interest in cases of negative RT-PCR [29]. High levels of IgM and IgG can be detected from the second week of symptom's onset, since IgM can be positive from the fourth day and IgG after 8 days. [30,31]. The accuracy of some serological tests is near 100% when samples are taken 20 days after infection [31].

Due to the limitations of detection kits available, a Covid-19 disease for the 28 cases of cohort #1.1 cannot be excluded, but the hypothesis that these deaths could be Covid-19 related is not evidence-based. According to literature [17,32], negative RT-PCR but typical chest CT features can be highly suggestive of Covid-19, particularly in areas with high prevalence of Covid-19. In fact, chest CT findings may be a more reliable, practical, and rapid method to assess a Covid-19 disease [12] and could have a role especially in cases with negative swabs. CT findings suggestive of Covid-19 were also observed in patients with negative RT-PCR results along with specific clinical symptoms [12].

Small-scale studies have demonstrated that the current RT-PCR testing has limited sensitivity [12], while chest CT may reveal pulmonary abnormalities consistent with Covid-19 in patients with initial negative RT-PCR results [17,33,34].

According to international radiological guidelines [22], these radiological abnormalities are mostly represented by GGOs, particularly on peripheral and lower lobes with/without crazy paving sign and multifocal patchy consolidation. They were found to be a typical radiological pattern that could be used to diagnose Covid-19 infection in patients with high clinical suspicion and negative initial RT-PCR [17,18]. CT abnormalities suggesting Covid-19 infection have been observed in up to 29.4% of patients with initially negative RT-PCR [26].

These radiological abnormalities are characterized by an increased lung opacity commonly found in Covid-19 related deaths, but they cannot be considered specific for Covid-19, as they can be also associated with other viral and bacterial infections. Non-typical CT findings have included pleural effusion, masses cavitation, and lymphadenopathies [23]. In fact, accumulation of pleural fluid is not a specific disease but the sign of underlying pathology such as congestive failure, cancer, pulmonary embolism, and pneumonia due to causes other than Covid-19 [35]. Pleural effusion can occur in Covid-19 related pneumonia but it has been found to be uncommon in previous studies. It can be a possible radiological finding observed with disease progression, reported in 3–4% of patients with Covid-19 pneumonia [36,37]. This is the reason why it has been classified as a non-typical CT finding compared to GGOs and consolidation with peripheral distribution by international radiological guidelines [13,22].

In cohort #1.1, 18 cases out of 28 with negative RT-PCR showed typical CT findings suggesting probable Covid-19 infection [18] compared to atypical CT findings for Covid-19 pneumonia observed in 10 cases. For these latter 10 cases, it is worth mentioning that atypical chest CT findings do not exclude a Covid-19 infection, especially in the first three days, just because chest imaging may also lead to both false-negative and false-positive results [19].

According to the current state of the art, all the 28 patients with negative RT-PCR (cohort #1.1) were classified to be suffering of pneumonia not COVID-19 related, although clinical and radiological chest CT findings were suggestive for COVID-19 in 18 out 28 cases. It is our opinion that diagnosis of COVID-19 cannot be excluded for these 18 victims who tested negative for Covid-19. Serological platforms are actually very useful to reveal infections based on the SARS-Cov-2-specific antibody responses, and they could be of help for the Covid-19 diagnosis.

Therefore, the discrepancy between typical CT findings and RT-PCR results is not new in literature [17,34,38], and it could be related to factors affecting the performance of these investigations according to their sensitivity and timing of examinations [27]. Unfortunately, false-negative and false-positive results of RT-PCR and CT-scan can occur, especially in the early stages of the disease, and they still need to be better identified and investigated. In fact, RT-PCR allows the identification of the infection in the early phase, even if there is a period of a few days, called the "window" period, where the subject is negative [28,38]. False negative RT-PCR results can be related to inadequate viral material (both quality and volume) at the time of specimen collection or technical issues during nucleic acid extraction [28,34]. In a systematic review of 919 patients [39], CT scan can also provide atypical findings, as the greatest severity of imaging has been reported around 10 days after symptom onset in the early stage of the disease. Imaging features can also vary depending on the disease stage during follow-up and timing of scanning, drug interventions, immunity status, and patient's age [17].

The novel coronavirus can be particularly harmful to people with pre-existing chronic conditions, which are common, especially among the elderly. In fact, patients belonging to cohort #1.1 had an age ranged between 65 to 87 years old with a median age of 79 years. A similar median age (75 years) was found in patients with positive RT-PCR (study group #2) with an age range between 55 and 94 years. Both study samples (group #2 and cohort #1.1) showed severe comorbidity represented mostly by hypertension and other cardiovascular diseases (chronic heart failure, dilatative cardiomyopathy secondary to myocardial infarction, atrial fibrillation, etc.), COPD, and diabetes. According to literature, these pre-existing conditions are relatively common in Covid-19 related deaths, and they correlate with poor clinical outcomes [4]. They may be serious risk factors for severe patients among which, in particular, hypertension and other heart disease pathologies are usually found in elderly population [40,41].

All causes of mortality can be interpreted as an approximation of the health status [20], and the full implications of the COVID-19 pandemic will be understood when more information on its pathogenesis and mechanisms of death is available. Due to the limited information available, most issues related to the dynamics of the COVID-19 disease have still significant uncertainties [42].

According to the definition for deaths due to COVID-19 provided by WHO in its guidelines [43], "a death due to Covid-19 should be resulting from a clinically compatible illness, in a probable or confirmed Covid-19 case, unless there is a clear alternative cause of death not related to Covid-19". Therefore, the cohort of 28 patients (cohort #1.1) suffering of pneumonia but with inconclusive testing for SARS-Cov-2 could be considered at least suspect cases, especially in cases with chest CT typical findings for COVID-19. Thus, false negative multiple RT-PCR results can be diagnostically challenging [17]. It is our suspect that Covid-19 mortality is currently underestimated due to the high rate of false negatives in the population tested for Covid-19, the lack of sensitivity of RT-PCR specimens, and factors affecting RT-PCR results (e.g., inappropriate sampling from upper or lower respiratory tract, inappropriate timing of sampling with a delay between administering and processing the test, low performance of detection kits) [28]. It is also worth mentioning that, in the early stage of disease, chest imaging can also provide atypical findings so that international radiological guidelines from the Centers for Disease Control (CDC) [44], the American College of Radiology (ACR) [45], and the British Society of Thoracic Imaging (BSTI) [22] do not recommend CT scans to diagnose Covid-19.

5. Conclusions

This study has several limitations, including small samples size and the fact that it focuses on deaths and not on all cases of Covid-19. However, according to literature, doubts can be raised about the accuracy of confirmed case and death counts, indicating a substantial underestimation of the magnitude of the burden of Covid-19 [20]. The excess of mortality could be higher than the one reported in the official epidemiological surveys. The true death rate from Covid-19 is strongly limited by the lack of reliable testing methods and full knowledge of this disease [42]. False negative cases can have a distorting effect on the assessment of real mortality rate and excess mortality. Furthermore, many who died from Covid-19 were likely never tested or they had false negative RT-PCR results or false negative chest imaging in the first three days. Other victims probably died from causes indirectly related to Covid-19 [20] and in these cases, autopsy can represent a reliable tool in the differential diagnosis [27].

Author Contributions: Conceptualization, P.M., C.S. and S.L.; methodology, P.Z., A.F., B.D.P. and C.P.C.; writing—original draft preparation, P.M., A.F. and C.S.; writing—review and editing, P.Z., B.D.P. and C.P.C.; supervision, S.L. and C.P.C. All authors have read and agreed to the published version of the manuscript.

Funding: This research received no external funding.

Institutional Review Board Statement: Data collection and analysis were approved by the Medical Ethics Committee of the "San Giuseppe Moscati" Hospital in accordance with international and institutional ethics guidelines as well as Italian legislation dealing with data protection.

Informed Consent Statement: Not applicable.

Data Availability Statement: The data presented in this study are available on request from the corresponding author. The data are not publicly available due to privacy restriction.

Conflicts of Interest: The authors declare no conflict of interest.

References

1. Wang, C.; Horby, P.W.; Hayden, F.G.; Gao, G.F. A novel coronavirus outbreak of global health concern. *Lancet* **2020**, *395*, 470–473. [CrossRef]
2. World Health Organization. *Clinical Management of Severe Acute Respiratory Infection When Novel Coronavirus (2019-nCoV) Infection Is Suspected;* WHO: Geneva, Switzerland, 2020. Available online: https://www.who.int/docs/default-source/coronaviruse/clinicamanagement-of-novel-cov.pdf (accessed on 16 February 2020).
3. Lu, R.; Zhao, X.; Li, J.; Niu, P.; Yang, B.; Wu, H.; Wang, W.; Song, H.; Huang, B.; Zhu, N.; et al. Genomic characterization and epidemiology of 2019 novel coronavirus: Implications for virus origins and receptor binding. *Lancet* **2020**, *395*, 565–574. [CrossRef]
4. Guan, W.J.; Liang, W.H.; Zhao, Y.; Liang, H.R.; Chen, Z.S.; Li, Y.M.; Liu, X.Q.; Chen, R.C.; Tang, C.L.; Wang, T.; et al. Comorbidity and its impact on 1590 patients with COVID-19 in China: A nationwide analysis. *Eur. Respir. J.* **2020**, *55*, 2000547. [CrossRef]
5. Ren, L.L.; Wang, Y.M.; Wu, Z.Q.; Xiang, Z.C.; Guo, L.; Xu, T.; Jiang, Y.Z.; Xiong, Y.; Li, Y.J.; Li, X.W.; et al. Identification of a novel coronavirus causing severe pneumonia in human: A descriptive study. *Chin. Med. J.* **2020**, *133*, 1015–1024. [CrossRef] [PubMed]
6. Huang, C.; Wang, Y.; Li, X.; Ren, L.; Zhao, J.; Hu, Y.; Zhang, L.; Fan, G.; Xu, J.; Gu, X.; et al. Clinical features of patients infected with 2019 novel coronavirus in Wuhan, China. *Lancet* **2020**, *395*, 497–506. [CrossRef]
7. Lu, H. Drug treatment options for the 2019-new coronavirus (2019-nCoV). *Biosci. Trends* **2020**, *14*, 69–71. [CrossRef] [PubMed]
8. Wang, W.; Tang, J.; Wei, F. Updated understanding of the outbreak of 2019 novel coronavirus (2019-nCoV) in Wuhan, China. *J. Med. Virol.* **2020**, *92*, 441–447. [CrossRef]
9. Li, Q.; Guan, X.; Wu, P.; Wang, X.; Zhou, L.; Tong, Y.; Ren, R.; Leung, K.S.M.; Lau, E.H.Y.; Wong, J.Y.; et al. Early Transmission Dynamics in Wuhan, China, of Novel Coronavirus-Infected Pneumonia. *N. Engl. J. Med.* **2020**, *382*, 1199–1207. [CrossRef]
10. Gandhi, M.; Yokoe, D.S.; Havlir, D.V. Asymptomatic Transmission, the Achilles' Heel of Current Strategies to Control Covid-19. *N. Engl. J. Med.* **2020**, *382*, 2158–2160. [CrossRef]

11. Baud, D.; Qi, X.; Nielsen-Saines, K.; Musso, D.; Pomar, L.; Favre, G. Real estimates of mortality following COVID-19 infection. *Lancet Infect. Dis.* **2020**, *20*, 773. [CrossRef]
12. Ai, T.; Yang, Z.; Hou, H.; Zhan, C.; Chen, C.; Lv, W.; Tao, Q.; Sun, Z.; Xia, L. Correlation of Chest CT and RT-PCR Testing for Coronavirus Disease 2019 (COVID-19) in China: A Report of 1014 Cases. *Radiology* **2020**, *296*, E32–E40. [CrossRef] [PubMed]
13. Revel, M.P.; Parkar, A.P.; Prosch, H.; Silva, M.; Sverzellati, N.; Gleeson, F.; Brady, A. European Society of Radiology (ESR) and the European Society of Thoracic Imaging (ESTI). COVID-19 patients and the radiology department-advice from the European Society of Radiology (ESR) and the European Society of Thoracic Imaging (ESTI). *Eur. Radiol.* **2020**, *30*, 4903–4909. [CrossRef] [PubMed]
14. Yang, Y.; Yang, M.; Shen, C.; Wang, F.; Yuan, J.; Li, J.; Zhang, M.; Wang, Z.; Xing, L.; Wei, J. Evaluating the accuracy of different respiratory specimens in the laboratory diagnosis and monitoring the viral shedding of 2019-nCoV infections. *MedRxiv* **2020**. [CrossRef]
15. Winichakoon, P.; Chaiwarith, R.; Liwsrisakun, C.; Salee, P.; Goonna, A.; Limsukon, A.; Kaewpoowat, Q. Negative Nasopharyngeal and Oropharyngeal Swabs Do Not Rule Out COVID-19. *J. Clin. Microbiol.* **2020**, *58*, e00297-20. [CrossRef] [PubMed]
16. Chung, M.; Bernheim, A.; Mei, X.; Zhang, N.; Huang, M.; Zeng, X.; Cui, J.; Xu, W.; Yang, Y.; Fayad, Z.A.; et al. CT Imaging Features of 2019 Novel Coronavirus (2019-nCoV). *Radiology* **2020**, *295*, 202–207. [CrossRef] [PubMed]
17. Brogna, B.; Bignardi, E.; Brogna, C.; Alberigo, M.; Grappone, M.; Megliola, A.; Salvatore, P.; Fontanella, G.; Mazza, E.M.; Musto, L. Typical CT findings of COVID-19 pneumonia in patients presenting with repetitive negative RT-PCR. *Radiography* **2020**. [CrossRef] [PubMed]
18. Caruso, D.; Polidori, T.; Guido, G.; Nicolai, M.; Bracci, B.; Cremona, A.; Zerunian, M.; Polici, M.; Pucciarelli, F.; Rucci, C.; et al. Typical and atypical COVID-19 computed tomography findings. *World J. Clin. Cases* **2020**, *8*, 3177–3187. [CrossRef] [PubMed]
19. Kwee, T.C.; Kwee, R.M. Chest CT in COVID-19: What the Radiologist Needs to Know. *Radiographics* **2020**, *40*, 1848–1865. [CrossRef]
20. Piccininni, M.; Rohmann, J.L.; Foresti, L.; Lurani, C.; Kurth, T. Use of all causes mortality to quantify the consequences of Covid-19 in Nembro, Lombardy: Descriptive study. *BMJ* **2020**, *369*, m1835. [CrossRef]
21. Impatto Dell'epidemia Covid-19 Sulla Mortalità Totale Della Popolazione Residente Periodo Gennaio-Maggio. 2020. Available online: https://www.istat.it/it/files/2020/07/Rapp_Istat_Iss_9luglio.pdf (accessed on 2 January 2021).
22. Nair, A.; Rodrigues, J.; Hare, S.; Edey, A.; Devaraj, A.; Jacob, J.; Johnstone, A.; McStay, R.; Denton, E.; Robinson, G. A British Society of Thoracic Imaging statement: Considerations in designing local imaging diagnostic algorithms for the COVID-19 pandemic. *Clin. Radiol.* **2020**, *75*, 329–334. [CrossRef]
23. Hani, C.; Trieu, N.H.; Saab, I.; Dangeard, S.; Bennani, S.; Chassagnon, G.; Revel, M.P. COVID-19 pneumonia: A review of typical CT findings and differential diagnosis. *Diagn. Interv. Imaging* **2020**, *101*, 263–268. [CrossRef] [PubMed]
24. Padhye, N.S. Reconstructed diagnostic sensitivity and specificity of the RT-PCR test for COVID-19. *medRxiv* **2020**. [CrossRef]
25. Onder, G.; Rezza, G.; Brusaferro, S. Case-Fatality Rate and Characteristics of Patients Dying in Relation to COVID-19 in Italy. *JAMA* **2020**, *323*, 1775–1776. [CrossRef]
26. Fang, Y.; Zhang, H.; Xie, J.; Lin, M.; Ying, L.; Pang, P.; Ji, W. Sensitivity of Chest CT for COVID-19: Comparison to RT-PCR. *Radiology* **2020**, *296*, E115–E117. [CrossRef] [PubMed]
27. Cipolloni, L.; Sessa, F.; Bertozzi, G.; Baldari, B.; Cantatore, S.; Testi, R.; D'Errico, S.; Di Mizio, G.; Asmundo, A.; Castorina, S.; et al. Preliminary Post-Mortem COVID-19 Evidence of Endothelial Injury and Factor VIII Hyperexpression. *Diagnostics* **2020**, *10*, 575. [CrossRef]
28. Sessa, F.; Bertozzi, G.; Cipolloni, L.; Baldari, B.; Cantatore, S.; D'Errico, S.; Di Mizio, G.; Asmundo, A.; Castorina, S.; Salerno, M.; et al. Clinical-Forensic Autopsy Findings to Defeat COVID-19 Disease: A Literature Review. *J. Clin. Med.* **2020**, *9*, 2026. [CrossRef]
29. Sethuraman, N.; Jeremiah, S.S.; Ryo, A. Interpreting Diagnostic Tests for SARS-CoV-2. *JAMA* **2020**, *323*, 2249–2251. [CrossRef]
30. Li, Z.; Yi, Y.; Luo, X.; Xiong, N.; Liu, Y.; Li, S.; Sun, R.; Wang, Y.; Hu, B.; Chen, W.; et al. Development and clinical application of a rapid IgM-IgG combined antibody test for SARS-CoV-2 infection diagnosis. *J. Med. Virol.* **2020**, preprint. [CrossRef]
31. Weissleder, R.; Lee, H.; Ko, J.; Pittet, M.J. COVID-19 diagnostics in context. *Sci. Transl. Med.* **2020**, *12*, eabc1931. [CrossRef]
32. Di Paolo, M.; Iacovelli, A.; Olmati, F.; Menichini, I.; Oliva, A.; Carnevalini, M.; Graziani, E.; Mastroianni, C.M.; Palange, P. False-negative RT-PCR in SARS-CoV-2 disease: Experience from an Italian COVID-19 unit. *ERJ Open Res.* **2020**, *6*. [CrossRef]
33. Huang, P.; Liu, T.; Huang, L.; Liu, H.; Lei, M.; Xu, W.; Hu, X.; Chen, J.; Liu, B. Use of Chest CT in Combination with Negative RT-PCR Assay for the 2019 Novel Coronavirus but High Clinical Suspicion. *Radiology* **2020**, *295*, 22–23. [CrossRef] [PubMed]
34. Xie, X.; Zhong, Z.; Zhao, W.; Zheng, C.; Wang, F.; Liu, J. Chest CT for Typical Coronavirus Disease 2019 (COVID-19) Pneumonia: Relationship to Negative RT-PCR Testing. *Radiology* **2020**, *296*, E41–E45. [CrossRef] [PubMed]
35. Jany, B.; Welte, T. Pleural Effusion in Adults-Etiology, Diagnosis, and Treatment. *Dtsch. Aerzteblatt Int.* **2019**, *116*, 377–386. [CrossRef]
36. Bai, H.X.; Hsieh, B.; Xiong, Z.; Halsey, K.; Choi, J.W.; Tran, T.M.L.; Pan, I.; Shi, L.-B.; Wang, D.-C.; Mei, J.; et al. Performance of Radiologists in Differentiating COVID-19 from Non-COVID-19 Viral Pneumonia at Chest CT. *Radiology* **2020**, *296*, E46–E54. [CrossRef]
37. Wong, H.Y.F.; Lam, H.Y.S.; Fong, A.H.; Leung, S.T.; Chin, T.W.; Lo, C.S.Y.; Lui, M.M.; Lee, J.C.Y.; Chiu, K.W.; Chung, T.W.; et al. Frequency and Distribution of Chest Radiographic Findings in Patients Positive for COVID-19. *Radiology* **2020**, *296*, E72–E78. [CrossRef]

38. Shen, M.; Zhou, Y.; Ye, J.; Al-Maskri, A.A.A.; Kang, Y.; Zeng, S.; Cai, S. Recent advances and perspectives of nucleic acid detection for coronavirus. *J. Pharm. Anal.* **2020**, *10*, 97–101. [CrossRef] [PubMed]
39. Salehi, S.; Abedi, A.; Balakrishnan, S.; Gholamrezanezhad, A. Coronavirus Disease 2019 (COVID-19): A Systematic Review of Imaging Findings in 919 Patients. *AJR Am. J. Roentgenol.* **2020**, *215*, 87–93. [CrossRef]
40. Yang, J.; Zheng, Y.; Gou, X.; Pu, K.; Chen, Z.; Guo, Q.; Ji, R.; Wang, H.; Wang, Y.; Zhou, Y. Prevalence of comorbidities and its effects in patients infected with SARS-CoV-2: A systematic review and meta-analysis. *Int. J. Infect. Dis.* **2020**, *94*, 91–95. [CrossRef] [PubMed]
41. Campobasso, C.P.; Bugelli, V.; De Micco, F.; Monticelli, F. Sudden cardiac death in elderly: The post-mortem examination of senile myocardium and myocardial infarction. *J. Gerontol. Geriatr.* **2017**, *65*, 223–231.
42. Lachmann, A. Correcting under-reported COVID-19 case numbers. *medRxiv* **2020**. [CrossRef]
43. International Guidelines for Certification and Classification (Coding) of Covid-19 as Cause of Death. Available online: https://www.who.int/classifications/icd/Guidelines_Cause_of_Death_COVID-19.pdf?ua=1 (accessed on 20 January 2021).
44. Interim Clinical Guidance for Management of Patients with Confirmed Coronavirus Disease (COVID-19). Available online: https://www.cdc.gov/coronavirus/2019-ncov/hcp/clinical-guidance-management-patients.html (accessed on 20 January 2021).
45. ACR Recommendations for the Use of Chest Radiography and Computed Tomography (CT) for Suspected COVID-19 Infection. Available online: https://www.acr.org/Advocacy-and-Economics/ACR-Position-Statements/Recommendations-for-Chest-Radiography-and-CT-for-Suspected-COVID19-Infection (accessed on 11 January 2021).

MDPI

Review

Adverse Effects of Anabolic-Androgenic Steroids: A Literature Review

Giuseppe Davide Albano [1,†], Francesco Amico [1,†], Giuseppe Cocimano [1], Aldo Liberto [1], Francesca Maglietta [2], Massimiliano Esposito [1], Giuseppe Li Rosi [3], Nunzio Di Nunno [4], Monica Salerno [1,‡] and Angelo Montana [1,*,‡]

1 Legal Medicine, Department of Medical, Surgical and Advanced Technologies, "G.F. Ingrassia", University of Catania, 95123 Catania, Italy; gdavidealbano@gmail.com (G.D.A.); francesco.amico@you.unipa.it (F.A.); giuseppe.cocimano@you.unipa.it (G.C.); aldoliberto@gmail.com (A.L.); massimiliano.esposito@you.unipa.it (M.E.); monica.salerno@unict.it (M.S.)
2 Department of Clinical and Experimental Medicine, University of Foggia, 71122 Foggia, Italy; francesca.maglietta@unifg.it
3 Department of Law, Criminology, Magna Graecia University of Catanzaro, 88100 Catanzaro, Italy; lirosigiose@gmail.com
4 Department of History, Society and Studies on Humanity, University of Salento, 73100 Lecce, Italy; nunzio.dinunno@unisalento.it
* Correspondence: angelo.montana@uniroma1.it; Tel.: +39-095-3782153
† These authors contributed equally to this work.
‡ These authors share last authorship.

Abstract: Anabolic-androgenic steroids (AASs) are a large group of molecules including endogenously produced androgens, such as testosterone, as well as synthetically manufactured derivatives. AAS use is widespread due to their ability to improve muscle growth for aesthetic purposes and athletes' performance, minimizing androgenic effects. AAS use is very popular and 1–3% of US inhabitants have been estimated to be AAS users. However, AASs have side effects, involving all organs, tissues and body functions, especially long-term toxicity involving the cardiovascular system and the reproductive system, thereby, their abuse is considered a public health issue. The aim of the proposed review is to highlight the most recent evidence regarding the mechanisms of action of AASs and their unwanted effects on organs and lifestyle, as well as suggesting that AAS misuse and abuse lead to adverse effects in all body tissues and organs. Oxidative stress, apoptosis, and protein synthesis alteration are common mechanisms involved in AAS-related damage in the whole body. The cardiovascular system and the reproductive system are the most frequently involved apparatuses. Epidemiology as well as the molecular and pathological mechanisms involved in the neuropsychiatric side-effects of AAS abuse are still unclear, further research is needed in this field. In addition, diagnostically reliable tests for AAS abuse should be standardized. In this regard, to prevent the use of AASs, public health measures in all settings are crucial. These measures consist of improved knowledge among healthcare workers, proper doping screening tests, educational interventions, and updated legislation.

Keywords: AASs; anabolic androgenic steroids; organ damage; toxicity; injury; chronic administration

Citation: Albano, G.D.; Amico, F.; Cocimano, G.; Liberto, A.; Maglietta, F.; Esposito, M.; Rosi, G.L.; Di Nunno, N.; Salerno, M.; Montana, A. Adverse Effects of Anabolic-Androgenic Steroids: A Literature Review. *Healthcare* **2021**, *9*, 97. https://doi.org/10.3390/healthcare9010097

Received: 5 December 2020
Accepted: 11 January 2021
Published: 19 January 2021

1. Introduction

Anabolic-androgenic steroids (AASs), commonly known as anabolic steroids, are a large group of molecules including endogenously produced androgens, such as testosterone, as well as synthetically manufactured derivatives [1]. Testosterone, Nandrolone Decanoate (ND), methandienone, and methenolol, are the most commonly abused androgens [2]. AAS use is widespread due to their ability to improve muscle growth for esthetic purposes and athletes' performance, minimizing androgenic effects [3]. Indeed, androgens are able to increase the size of muscle fibers as well as muscle strength, and while their use was initially restricted to professional bodybuilders, nowadays it has become more popular among recreational athletes [4,5]. AAS anabolic properties have been widely used for therapeutic purposes. Indeed, AASs had a role in the treatment of chronic kidney disease and osteoporosis in postmenopausal women, as well as inoperable breast cancer, and for diseases characterized by a negative nitrogen balance [2]. However, use of AASs is forbidden by the World Anti-Doping Agency (WADA). However, AAS use is still very popular and 1–3% of US inhabitants have been estimated to be AAS users [6]. Moreover, in younger subjects' higher estimates have been reported [7,8]. However, AASs have side effects involving all organs, tissues and body functions, especially long-term toxicity involving the cardiovascular system and the reproductive system, therefore, their abuse is considered a public health issue [9,10]. In this regard, an increased awareness is needed among the population and healthcare workers, both for diagnostic, therapeutic and prevention purposes. The aim of the proposed review is to highlight the state of the art regarding the mechanisms of action of AASs and the adverse effects related to AAS use/abuse.

2. Phisiology of AASs

The anabolic androgenic effects are related to the androgen receptor (AR)-signaling action. Androgen receptors are widespread in human tissues and organs. There are three main action mechanisms: (i) direct binding to androgen receptor; (ii) via dihydrotestosterone (DHT) produced by the action of 5- a-reductase, and (iii) via estrogen receptors by means of estradiol produced by CYP19 aromatase. In particular, free testosterone is transported into target tissue cell cytoplasm; binding to the AR takes place either directly or after conversion to 5αdihydrotestosterone (DHT) by the cytoplasmic enzyme 5-alpha reductase. Into the cell nucleus, both free or bound, testosterone binds specific nucleotide sequences of the chromosomal DNA. The produced DNA activate the transcription of specific responsive genes, with significant influence on protein synthesis [11–13]. After dimerization the complex binds to specific promoter areas of target genes called androgen response elements (AREs), influencing the transcription process [14]. Furthermore, non-genomic pathways, by interfering with the G-protein coupled receptor, a transmembrane receptor located inside the cell, can lead to rapid steroid hormone activation [6,15]. In this regard, sex steroids might influence thyroid function as a consequence of the expression of androgen receptors in this tissue, leading to thyrocyte proliferation in culture independently from TSH [16]. The same mechanism has been described in other tissues [17].

3. Pathophysiology of AASs

The most relevant mechanisms that lead to the increase of AASs in circulation are: administration of testosterone or its synthetic derivatives or administration of drugs that raise endogenous testosterone production [11]. The mechanism of action of AASs in supraphysiological doses is characterized by the impairment of testosterone biosynthesis in tissues (Figure 1).

AASs exert their effects by activating androgen receptor (AR) signaling. Several parts of the body are involved because of the presence of ARs in many tissues [12]. At normal physiologic levels of testosterone androgen receptors are saturated and the AASs effects may be a consequence of other mechanisms rather than androgen receptors activation. High testosterone levels may have an antagonist effect on glucocorticoid receptors, leading to inhibition of glucose synthesis and protein catabolism. Indeed, high dose AASs may displace glucocorticoids from their receptors, decrease proteins breakdown in muscles, leading to an increase in muscle mass and muscle strength [18]. The inhibition of glucocorticoid action is also due to the stimulation of growth hormone (GH) and insulin-like growth factor (IGF)-1 axis. In this regard, AASs induce an androgen-mediated stimulation of GH and the hepatic synthesis of IGF-1, leading to muscle proteins formation and anabolic effects [5]. Moreover, testosterone is converted by aromatase action to estradiol and estrone, influencing brain and sexual differentiation, bone and muscle mass increase, puberty and sexual functions. High doses of AASs exert an antiestrogenic effect due to a down-regulation of androgen receptors and a competition with estrogens with their receptors [18].

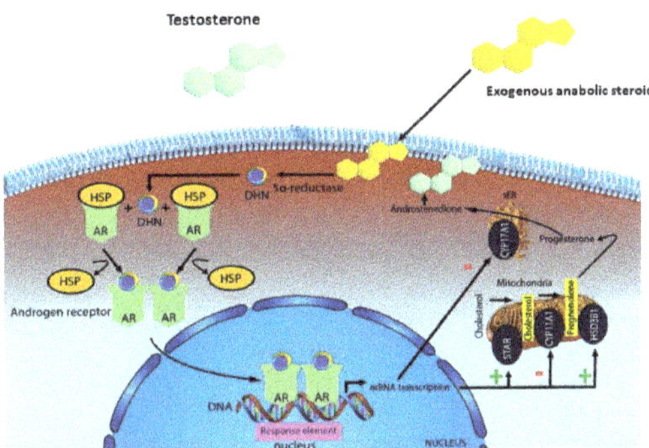

Figure 1. Mechanism of action of exogenous anabolic steroids: an anabolic steroid is transported into the target tissue cell cytoplasm where it can either bind the androgen receptor, or be reduced by the cytoplasmic enzyme 5-alpha reductase. The N-receptor complex undergoes a structural change that allows its translocation into the cell nucleus, where it directly binds to specific nucleotide sequences of the chromosomal DNA. The produced DNA interferes with the physiological biosynthesis of testosterone.

Thereby, AASs effects are the result of the amplification of testosterone and estrogens physiologic consequences. Several experimental human studies showed the influence of testosterone and AASs doses concentration on their effects. According to a double-blind human study, low dose administration of methyltestosterone is considered 40 mg/d and high dose is 240 mg/d [19]. In this study, 3 days' administration of high doses of methyltestosterone led to neuropsychiatric effects. Another study found psychological effects after 14 weeks of 500 mg administration of testosterone cypionate for week [18]. Moreover, a recent study found that in a population of AASs users the weekly dose assumption ranged 75–1550 mg/week [20].

Several studies have suggested an influence of AASs on oxidative stress [18–22]. AASs may alter the function of mitochondrial respiratory chain complexes and mono-oxygenase systems, thereby causing an imbalance between free radical production and their subsequent amelioration [18]. A recent study conducted on Wistar rats treated with ND demonstrated a disruption of the redox metabolism in the animals' tissues through the increase of reactive oxygen species (ROS) that may play a role in DNA damage [19]. In this regard, the beneficial effect of physical activity in diminishing oxidative stress as a consequence of the upregulation of antioxidant enzymes is well known. However, a role of ND in physical activity-induced cardio-protection impairment has been demonstrated, therefore AASs may play a role in cardiac ischemic injury mediated by oxidative stress [21]. Such data have also been confirmed in a human study that investigated the effects of a supraphysiological administration of testosterone enanthate (500 mg) in healthy volunteers. In this study an impairment of endothelial function as a consequence of the dysfunction of antioxidative capacity following testosterone administration was demonstrated [22]. Furthermore, oxidative stress plays a leading role in AAS-mediated neurotoxicity: androgens may be neuroprotective in cases of low levels of oxidative stress, however, they may increase brain damage in cases of elevated oxidative stress [23].

AAS-related damage is also associated with apoptosis activation [23–26]. Indeed, it was demonstrated that supraphysiological concentrations of AASs may induce neurotoxicity by involving the apoptotic process and neurodegeneration [24]. Moreover, AASs are responsible for excitotoxicity induced by N-methyl-d-aspartate (NMDA) in primary cultures of mouse cortical cells [24]. A recent study suggested how AASs may induce apoptosis and oxidative stress in the rat brain because of the activation of the non-genomic pathway and the elevation of the intracellular calcium concentration [25]. Another study confirmed a role of apoptosis in AAS-related damage. Androgens may increase the expression of profibrotic cytokines such as TGF-b in mice kidneys, leading to the activation of the apoptotic process and the promotion of focal segmental glomerulosclerosis. Moreover, environmental and inflammation stress can lead to proteotoxic damage and dysregulate heat shock proteins [26]. In this regard, an activation of the extrinsic pathway of the apoptosis in the vascular smooth muscle cells in rats treated with testosterone was observed [27].

AASs are characterized by the activation of protein synthesis. The supraphysiological administration of ND decreased the fat mass and increased the protein mass in treated Wistar rats due to amino acid uptake and protein synthesis amelioration [20]. A similar mechanism was described in skeletal muscles: increased protein synthesis, a decrease in protein breakdown, the elevated formation of new myotubes and myonuclei lead to an increase in muscle mass and strength as well as an increased exercise capacity [21]. Moreover, testosterone has anti-inflammatory effects and improves insulin sensitivity because of its capacity to reduce the expression of proinflammatory cytokines, such as interleukin-1β, interleukin-6, and reduce the circulation of inflammatory cells [21].

Recent clinical and experimental studies proved that the increased activity of the renin-angiotensin-aldosterone system (RAAS) plays a pivotal role in the pathogenesis of cardiological diseases. The over-activation of the RAAS may lead to changes in the cardiovascular system observed in subjects taking AASs for doping purposes. The pathogenesis of cardiological changes and many tragic cases resulting from AAS abuse could depend on the strong stimulation of the RAAS and enforced effects by the tissue aldosterone action. A raised level of aldosterone is considered to be related to the occurrence of cardiac illnesses, independently of increased blood pressure. Lastly, the results of a study demonstrated that the presence of AAS metabolites in urine may be a predictive factor of cardiac changes in AAS abusers [27]. Oxidative stress, apoptosis, inflammation and changes in endocrine homeostasis are responsible for multi-organ damage in AASs abusers. Although several positive effects of AAS use had been described, supraphysiological doses and AASs abuse and misuse may lead to serious consequences in all body tissues and organs (Figure 2).

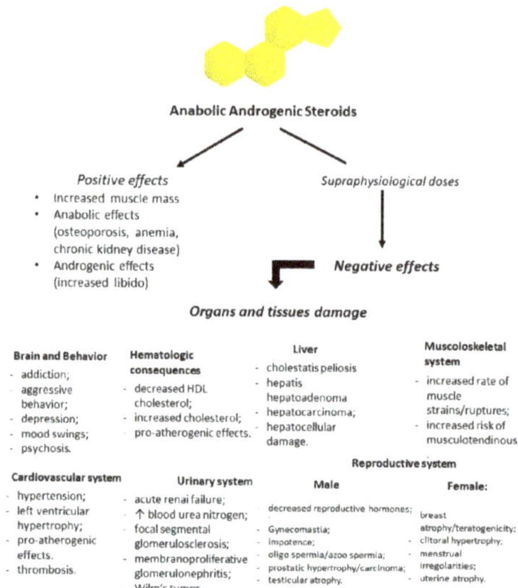

Figure 2. Flowchart of positive and negative effects of anabolic-androgenic steroid (AAS) administration. Prolonged and high doses of testosterone and his derivatives lead to serious consequences in all body tissues and organs.

4. AAS Use and Adverse Effects

Several mechanisms are involved in AAS adverse effects and need to be better clarified. AAS related effects involve several organs and systems, both in animals and humans. This is possibly due to the widespread presence of AR in the body and to the impairment of biosynthesis, transformation and degradation of endogenous steroids [28]. AASs bind to a specific type of androgen receptor and by the time the receptors are saturated, AASs in supraphysiological doses may lead to secondary effects [29–31]. However, side effects associated with AAS use (i.e., under medical supervision) have to be differentiated from those caused by abuse (i.e., consumption of many drugs at high doses; any nonmedical use of these substances) [32]. Some athletes consume multiple drugs in addition to anabolic steroids such as alcohol, opioids, cocaine, marijuana, and gamma hydroxybutyrate, some of which can interact adversely with AASs. Polydrug assumption makes it hard to attribute the observed effects to a single drug. AAS effects are also related to sex, dose and duration of administration. In this regard, most of the effects are observed after long-term administration [33].

4.1. Autopsy Findings

As we mentioned before the prolonged misuse and abuse of AASs can lead to several adverse effects, some of which may be even fatal especially the ones regarding the cardiovascular system, such as sudden cardiac death and coronary artery disease [34,35]. A recent autopsy series described that cardiovascular disease was widespread in AAS-related deaths [36]. Another series showed that all cases had the same findings: absence of asymmetrical left ventricular hypertrophy, coronary atherosclerosis causing significant luminal narrowing, pulmonary thromboembolism, coronary and endocavitary thrombi, and inflammatory infiltrates. Furthermore, the histopathologic study showed myocardial damage characterized by myocyte hypertrophy, focal myocyte damage with myofibrillar loss, interstitial fibrosis, mostly at the subepicardial, and small vessel disease [37]. Another study reviewed all the 19 AAS-related deaths cases presented in the literature, highlighting that in all cases extracardiac causes were excluded, except for one case regarding venous thromboembolism [35].

It was demonstrated that AASs increase the risk of premature death, especially among subjects with other pathologies and/or psychiatric diseases [36]. A survey conducted in 21 gyms in Britain reported that 8% of respondents declared having taken AASs in their life. Another study in the UK showed that 70% of the clientele in a health and fitness community were AAS users [37].

The toxicological investigation executed mostly on urine samples but also on blood and hair samples, by performing several screening tests and analytical methods, showed the presence of AASs and/or their metabolites in urine specimens in 12 cases; in one case nandrolone was detected in blood, while in another case stanozolol was detected in a hair sample [35]. Another study showed that 35% of the users examined were found to be positive for two or more AASs in connection with autopsy. Moreover, an association between the use of AASs and other illicit drugs, such as cannabis, cocaine, amphetamines or LSD, was observed. The combination of physical activity and prolonged/chronic or previous misuse of AASs leads to a predisposition to different patterns of myocardial injury and sudden cardiac death [35]. When performing an autopsy in a sudden death case involving a young athlete, attention to the physical phenotype such as muscular hypertrophy, striae in pectoral or biceps muscles, gynecomastia, testicular atrophy, and acne is mandatory in order to suggest AAS abuse and perform a detailed examination of the heart. Chemico-toxicological analysis is a crucial tool to assess the link between sudden cardiac death and AAS abuse [38]. Autopsy plays a pivotal role in the study of AAS adverse effects and organ damage related to their use/abuse. Moreover, autopsy studies may provide useful information regarding the pathophysiology of the effects of AAS long-term administration, therefore autopsy practice should be implemented in suspected AASs-related deaths.

4.2. Brain and Behavior

The neurotoxic action of AASs is associated with both membrane AR and G-protein coupled receptors [39]. Furthermore, several studies highlighted the role of apoptosis in determining brain damage [24,25,32,40,41]. Indeed, it was demonstrated that high concentrations of methandienone and 17-a-methyltestosterone provoke detrimental effects on neuron cell cultures expressing AR, inhibiting neurite network maintenance, leading to cell death through apoptosis and cleavage of protective chaperone proteins, such as Hsp90 [24].

A recent study suggested that miRNA dysregulation may be involved in the mechanisms that characterize AAS-related brain damage. In this study three groups were investigated: "AAS" users, "Cocaine" abusers and "Aging" people. In this regard, miR-34 and miR-132 were considerably higher in the "AAS" group [42].

The presence of apoptosis in brain areas of rats treated with long-term administration of nandrolone was suggested in a recent study. In this regard, a link between oxidative stress and NF-Kb signaling was described, promoting brain injury in specific areas, such as the hippocampus, striatum and frontal cortex [32]. Furthermore, it was found that daily injections of stanozol in male adult rats for 28 days led to histopathologic changes in the hippocampus by activating apoptotic and pre-apoptotic cells [40]. Moreover, another study demonstrated that supraphysiological doses of AASs impair the beneficial effects of physical activity on hippocampal cell proliferation and apoptotic signaling [41]. Endurance exercise improves the redox system balance by stabilizing the mitochondrial membrane, leading to a reduction of apoptotic effects of ND in neural cells [25].

Cognitive function may also be impaired by AAS abuse. Weightlifters exposed to AASs had lower cognitive functions, such as motor and executive performance, compared to nonexposed subjects [43]. According to a recent study that performed a neuroimaging investigation of AAS users, smaller overall gray matter, cortical and putamen volume, and thinner cortex in widespread regions in AAS users compared to non-using weightlifters was observed [44]. Furthermore, another imaging study showed markedly increased right amygdala volumes; markedly decreased right amygdala and reduced dACCgln/glu and scyllo-inositol levels compared to nonusers [45]. Recent evidence, by administrating neuropsychological tests to weightlifters both AAS users and nonusers, demonstrated a cognitive disfunction due to long-term high AAS exposure [46]. In this regard, oxidative stress and apoptosis due to AASs abuse may lead to neurodegeneration and dementia, especially in long-term users, adolescents and young adults [47,48].

AASs in supraphysiological concentrations influence several central nervous system functions, such as memory, aggressiveness, anxiety and depression, particularly in predisposed individuals [48–52]. The underlying mechanisms involve neurotransmission by affecting the synthesis and degradation of neurotransmitters, as well as neurotransmitter metabolism [53]. In addition, an animal study suggested that long-term administration of ND leads to anxiolytic behavior and memory impairment. [54,55]. Chronic administration of high doses of AASs is related to anxiety-like behavior through the corticotrophin release factor by enhancing GABAergic inhibitory effects from the central amygdala onto the bed nucleus of the stria terminalis [56]. Moreover, chronic AAS administration changes neurotransmitter expression involved in aggression control [57–59]. Lastly, AASs may induce NMDA receptor phosphorylation in order to increase excitatory neurotransmission, resulting in an increment of aggression [60]. In this regard, the orbitofrontal cortex may play a role in the aggressiveness and violent behavior due to AAS consumption. Indeed, the reduction of the orbitofrontal cortex observed in such cases may lead to the lack of inhibitory control [50]. People who use AASs have a higher probability to be drug and alcohol abusers [4,33,57]. Long-term research is needed to clarify the mechanisms and the organic and social processes involved in neuropsychiatric effects of AAS abuse.

4.3. Cardiovascular System

Notwithstanding the elevated morbidity and mortality, cardiac and metabolic consequences of AAS abuse are still unclear [59–62]. Cardiac injury is the most frequent consequence of the administration of exogenous steroids, due to its susceptibility to oxidative stress and its important metabolic activity, compared with the remaining body tissues and organs [63]. Chronic administration of high doses of AASs is responsible for the dysfunction in tonic cardiac autonomic regulation. Indeed, an experimental study demonstrated that rats treated with AASs were characterized by the impairment of parasympathetic cardiac modulation, decreased high frequency power and heart rate variability [64]. Furthermore, the inflammatory process may play a role in triggering cardiac injury in AAS abusers. In fact, in a mouse model a strong cytokine reaction was observed in mice treated with anabolic steroids compared to the control group, suggesting a role of TNF-α in determining myocardial injury [65]. Moreover, it was demonstrated that after administration of anabolic steroids treated animals lost the adaptive response of exercise-induced amelioration of antioxidant activity [21,32,58]. AAS use in supraphysiological doses is associated with abnormal plasma lipoproteins [59,66,67]. A human study including hypogonadal men undergoing substitutive therapy with testosterone showed decreased plasma levels of high-density lipoprotein (HDL) cholesterol [68]. Other studies found hyperomociysteinaemia and increased low-density lipoprotein (LDL) cholesterol levels after long-term AAS administration, underlining the promotion of atherogenesis of these substances [66–68]. An increased sympathetic activity was observed after AAS administration [69]. Moreover, high concentrations of AASs by activating ARs, cell membrane receptors and secondary transmitters stimulate the renin-angiotensin-aldosteron system, leading to an increased synthesis of heart muscle, left ventricular hypertrophy and hypertension [27]. AAS users show higher left ventricular mass index, thicker left ventricular walls, more concentric left geometry and myocardial mechanical dysfunction compared to non-users [70–72]. Long-term training associated with AAS administration reduce left ventricle relaxation properties [73]. In this regard, the use of anabolic steroids is associated with the loss of the beneficial effects on left ventricle function induced by exercise [74]. Arrhythmic events following long-term administration of AASs were reported [68,75–78]. Atrial fibrillation is the most frequent event but ventricular arrhythmias and sudden cardiac death were described in literature. Both human and animal studies showed an association between testosterone administration and the impairment of cardiac repolarization [76–79]. Moreover, cardiac hypertrophy induced by AASs plays an important role in electric and morphologic heart disturbances [77]. AAS users are characterized by an increased volume of atherosclerotic plaque [80]. In addition, 3% of AAS users had myocardial infarction as a consequence of atherosclerotic disease [81]. The mechanisms involved in AAS-induced myocardial infarction are the following: atherogenesis, thrombosis and vasospasm. Experimental data showed that in animal treated with AASs there were increased thrombotic stimuli. Furthermore, AAS abuse is related to endothelia dysfunction by impairing both endothelial-dependent and endothelial-independent vasodilatation [62]. AAS abuse raises the risk of life-threatening arrhythmias and sudden cardiac death. The hypothesized mechanisms are pro-arrhythmic effects of AASs, induction of myocardial ischemia, structural changes and repolarization impairment [78,79].

A recent study suggested that Doppler myocardial imaging is a useful tool to detect subclinical left ventricular dysfunction in AAS athlete abusers [79]. New imaging tools, such as magnetic resonance, may give fundamental information regarding myocardial tissue in these cases [76]. Clinicians must be aware of the mechanisms involved in cardiotoxicity, the pathological and clinical consequences, as well as the diagnostic tools to highlight cardiac damage in AAS abusers, in order to consider AAS abuse in differential diagnosis and undertake primary and secondary prevention in their patients.

4.4. Liver

Hepatotoxicity is one of the most frequent side effects of AAS abuse [82,83]. AAS-induced hepatotoxicity was hypothesized to be related to oxidative stress in hepatic cells. Following AR receptor activation an increase in reactive oxygen species can be observed due to the increase in mitochondrial b-oxidation. Moreover, antioxidant substances have a protective role against hepatotoxicity mediated by AASs. It was also demonstrated that androgenic potency and metabolic resistance are positively linked to the degree of liver damage [82].

AAS-induced hepatotoxicity is influenced by genetic factors, and is related to the infiltration of inflammatory cells in liver tissue, such as lymphocytes, neutrophils and eosinophils [83,84]. Oxidative stress could have a role in determining liver damage consequent to AAS abuse by activating androgen receptors that lead to mitochondrial degeneration of hepatic cells [84]. A recent study evaluated the liver effects of 5 weeks of administration of ND in rats. The results highlighted an increase of plasma levels of liver necrosis markers, an increase in collagen deposition in liver parenchyma, portal space, and the centrolobular vein [84]. The mechanism involved in collagen deposition could be the increase in the number and in the activity of Kuppfer cells. In this regard, Kuppfer cell activation leads to the production of many inflammatory cytokines such as TGF-b1, NF-Kb, IL-1b, related to the liver fibrosis process [85,86].

Two of the most common liver consequences following supraphysiological doses of AASs are peliosis and cholestasis. Peliosis is characterized by multiple blood-filled cavities histologically characterized by the presence of scattered, small, blood-filled cystic spaces throughout the liver parenchyma [83]. The mechanism involved could be the induction of hyperplasia of the hepatocytes responsible for mechanical obstruction of hepatic veins and the genesis of nodules and tumors [83–85]. In addition, animal studies demonstrated that bile accumulation can be a consequence of the reduction of his transportation ability. However, AAS-associated cholestasis is not characterized by the presence of necrosis and inflammation. Inflammation and necrosis may lead to a regenerative signal in AAS-induced hepatotoxicity [83,87]. A correlation between hepatocellular adenoma and androgen steroid therapy was described in the literature and the risk of androgen-associated liver tumor seems to be related to the dose and the potency of AAS administration [83,84]. Further studies are warranted to clarify the correlation between AASs administration and hepatocellular proliferation, in order to undertake preventive measures.

4.5. Urinary System

Several studies highlighted that prolonged androgen exposure has a direct toxic effect on kidneys, especially glomerular cells, causing accumulation of mesangial matrix, podocyte depletion and structural adaptations [26,87–89]. In this regard, kidney tissues are characterized by the expression of ARs. AR activation leads to cell growth and hypertrophy in the kidney. A recent report suggested that ND exposure promotes hypertrophy in proximal and distal convoluted tubules of mice kidneys [90]. Moreover, both testosterone activity and direct ND action to AR may play a role in the genesis of kidney fibrosis after long-term ND exposure [89].

Prolonged administration of ND in mice has been shown to cause dose-dependent oxidative kidney stress and damage. Mice kidneys treated with ND exhibited increased lipid peroxidation and decrease antioxidant enzyme activity, such as glutathione reductase and glutathione peroxidase [87]. A recent study suggested a dose related oxidative stress in mouse kidneys treated with prolonged doses of ND [87]. The authors observed an increase in lipid peroxidation markers and an increase of pro-inflammatory and pro-apoptotic markers such as IL-1B, Hsp90 and TNF associated with a decrease of antioxidant enzymes, which could lead to secondary focal segmental glomeruloscelerosis [87].

Morphological changes were observed in mice treated with ND. Three months after intramuscular injection of androgen, several histopathological alterations were detected: glomerular atrophy and fragmentation, tubular wall rupture, vacuolar degeneration of the epithelium lining of the proximal convoluted tubules and blood hemorrhage between the tubules, basal lamina thickening in distal convoluted tubules and tubes with only the basal lamina, many hyaline cylinders, some areas of necrosis, eosinophilic cell cytoplasm, which is a sign of chronicity and vascular congestion, were found in kidney samples [91]. As in other tissues and organs, oxidative stress, apoptosis and inflammation play a pivotal role in urinary system damage. This information is fundamental for therapeutic and prevention measures.

4.6. Muscoloskeletal System

Muscle mass seems to be influenced by AAS administration [30,92,93]. In fact, testosterone, by binding to AR, produces an increased production of IGF-1, a decreased expression of myostatin and the differentiation of pluripotent mesenchymal cells into a myogenic lineage. These mechanisms are involved in an increase in protein synthesis, a decrease in protein breakdown, the formation of new myotubes as well as the increase in myonuclei number, thereby leading to the increase in muscle mass, strength and exercise capacity [94].

In addition, high concentrations of AASs can provoke serious skeletal muscles injuries [95]. An experimental study demonstrated that supraphysiological doses of AASs induce a decrease in MMP-2 activity in the agonist jumping rat muscles [96]. It was suggested that the vascular endothelial growth factor (VEGF) may play a role in the mechanism involved in skeletal exercise adaptation. VEGF expression was reduced in rats who underwent ND administration and this is possibly related to MMP-2 activity dysfunction, since MMPs are involved in the regulation of VEGF extracellular stores [97]. Moreover, the decreased expression of VEGF may play a role in skeletal damage due to AASs, as a consequence of poor remodeling and poor vascularization [97]. Nevertheless, AASs could also be involved in tendon damage [98,99]. The morphology and the organization of collagen fibers may be modified by physical activity. In this regard, AAS abuse also increases the risk of tendon rupture, due to the increase of muscle mass, strength and the inability to respond, especially during exercise [98]. It was demonstrated that ND increased tendon remodeling despite decreases in MMP-2 activity in rat tendons [99]. However, AAS-related MMP dysregulation still needs to be better clarified. Esthetic purposes, increase of muscle mass and strength are one of the most frequent reason why young people and athletes are AASs abusers. Information campaign and public health measures are needed to increase the awareness in young population regarding muscoloskeletal side effects of AASs abuse.

4.7. Reproductive System

Androgens play a pivotal role in the development of male reproductive organs. They are necessary for puberty and male sexual function. AAS administration leads to a negative feedback on the hypothalamic-pituitary axis, altering the secretion of both FSH and LH, causing infertility [5]. A recent study focused on Leydig cell cultures treated with ND, demonstrating an impairment of testosterone production due to STARR and CYP17A1 expression interference in these cells [100].

Previous studies suggested that both current and past AAS users reported increased frequency of morning erections, sexual thoughts, and satisfaction. However, several side effects were observed such as erectile dysfunction, anorgasmia and premature ejaculation. Nevertheless, recent findings support the hypothesis that increased frequency and duration of high-dose AASs lead to sexual dysfunctions following discontinuation [101]. Prior AAS use is frequent among young men with hypogonadism and this element needs to be taken into consideration in the clinical evaluation of hypogonadism [102].

Animal histological studies of testes demonstrated spermatogenesis impairment with lack of advanced spermatidis and decreased number of spermatidis due to AAS use [5,88]. An impairment of the blood-testis-barrier was also observed in CD1 mice treated with ND, which may play a role in triggering the spermatogenesis alteration [103]. Moreover, quantitative changes in number, diameter and thickness of seminiferous tubules were detected in albino rats after AAS administration [104]. Apoptosis has been reported to play an important role in the regulation of germ cell populations in the adult testes. The correlation between apoptosis and high AAS doses and exercises has recently been experimentally assessed in animal models. Shokri et al. report a significant increase in the rate of apoptosis of spermatogenic cells after nandrolone administration, an increase clearly amplified by physical exercise [5].

According to the length of use of anabolic steroids and the period since the last drug administration prior to the survey, changes in sperm parameters were observed in a study: the percentages of motile sperm decreased among bodybuilders compared to healthy volunteers. Moreover, in relation to these results even after prolonged use of extremely high concentrations of anabolic steroids, sperm production can return to physiological rates for bodybuilders who stop consumption of anabolic steroids for 4 months [5]. Furthermore, some authors found alterations in sperm quantity, protamine, and DNA integrity in Wistar rats that underwent exercise treated with high concentrations of ND. The incidence of AASs in protamine and DNA fragmentation is a relevant issue in the study of male fertility, given that these drugs are used in high doses as in the case of some athletes. In addition, a direct relationship between irregular protamine expression and sperm count, motility, morphology, or fertilization was reported [104].

According to recent data, 20% of patients who were being treated for symptomatic hypogonadism had previously used AASs. Recognition of the specific details of the user's AAS exposure is crucial for their medical management. Management strategies for male infertility secondary to AAS induced hypogonadism should focus on the underlying hypogonadal state [105,106]. According to a recent study, chronic AAS abuse should be considered when a muscular man presents with hypogonadism, onset of gynecomastia or hirsutism.

4.8. Hematologic Consequences

Before the introduction of recombinant human erythropoietin, AASs were used in the treatment of anemias, indeed, AASs are capable of increasing erythropoietin secretion. Other AAS induced side effects are the increase of hematocrit and erythrocytosis [93]. AAS abuse has been recurrently associated with an increased risk of thrombosis and is detrimental to cardiovascular health [107–109]. However, the association has primarily been based on case reports. Increased LDL and decreased HDL are linked to an increased cardiovascular risk. Mild, but significant, increases in mean red blood cell, hematocrit, hemoglobin, and white blood cell concentrations in 33 men were described after intramuscular testosterone enanthate, 200 mg every 3 or 4 weeks for 24 weeks [93]. The influence of AASs on plasma concentration and function of coagulation factors depends on the substance and the dose of the anabolic steroid [110]. In this regard, it was demonstrated that physiological testosterone stimulates tissue plasminogen activator and tissue factor pathway inhibitor and inhibits plasminogen activator inhibitor type 1 release in endothelial cells. The relationship between AAS abuse and thrombosis has not been sufficiently clarified by the current literature of which only a few reports concern actual thrombotic outcomes [11]. A recent report suggested a possible correlation between AAS abuse and immunodeficiency that may be related to a mimicking action of corticosteroid activity. Moreover, this report suggested that AAS abuse should be investigated when an uncommon death occurs in immunosuppressed patients [111,112].

4.9. AASs and Cancer

The biochemical mechanism of AASs is similar to that of testosterone. AASs bind to DNA sequences and induce gene expression alterations. In a recent review regarding androgen effects on cellular functions, it was stated that a combination among genetic and epigenetic factors is the cause of toxicity, mutagenicity, genotoxicity and carcinogenicity of sexual hormones [113]. However, AAS related genotoxicity still remains unclear. Epigenetic molecular mechanisms, which lead to a genetic transcription control are: DNA methylation, histone modifications and chromatin condensation [114]. DNA methylation inhibits the binding between transcriptional factors and their target sequences, both promoters and introns, preventing transcriptional expression activation. Chromatin condensation also regulates transcriptional expression [115,116]. Testosterone synthetic derivatives can be metabolized, in adipose, cerebral and testicular tissues, to 17β-estradiol, a known potentially mutagenic and carcinogenic steroid [113]. 17 beta-estradiol and its metabolites are also considered inducers of cell proliferation. Furthermore, during their catabolism, AASs reveal their oxidative role, increase reactive oxygen species (ROS) production, which are highly unstable, easily lose hydrogen atoms, form covalent bonds with DNA bases or sequences, and may induce genetic damage [113].

It has been suggested that the incidence of cancer in different tissues is strictly positively correlated to the number of stem cell divisions in the lifetime occurring in them [50]. On this basis, it can be hypothesized that the chronic administration of nandrolone, favoring the persistence and viability of stem cells in different tissues, could represent a preconditioning that, in addition to multiple hits, could enhance the risk of carcinogenesis onset especially in stem cell-rich tissues such as the liver [117,118]. The side effects on the natural synthesis of anabolic steroids play a potential role in hormonal changes/regulation and it could be suspected to be at the base of certain carcinogenic mechanisms [113,119]. Furthermore, easily accessible and commonly diffused AASs, such as nandrolone and stanozolol, have the potential role in the pathogenesis of cancer, such as Leydig cell tumor through multiple process pathways [113]. Given that it was demonstrated a correlation between AASs abuse and cancer, the prevention of its abuse and the information campaigns in gyms and among young athletes are mandatory. In this regard, surveillance of long-term abuser is warranted in order to perform at early diagnosis.

Adverse effects of anabolic steroid use are summarized in Table 1.

Table 1. Summary of adverse effects of anabolic steroid use.

Organ/System	Author(s)	Year of Publication	Adverse Effects
Brain and Behavior	Bertozzi, et al.	2019	↓ orbitofrontal cortex; lack of inhibitory control.
	Hauger, et al.	2019	↓ memory, ↓ anxiety ↑ depression.
	Joksimovic, et al.	2019	↓ orbitofrontal cortex; lack of inhibitory control.
	Karimooy, et al.	2019	Neurodegeneration; histopathologic changes in hippocampus.
	Bertozzi, et al.	2018	↑ aggressiveness.
	Bjørnebekk, et al.	2017	Lower cognitive functions, motor and executive performance; ↓ gray matter, cortical and putamen volume.
	Bueno, et al.	2017	Neurodegeneration.
	Turillazzi, et al.	2016	Neurodegeneration; hippocampus, striatum and frontal cortex injury.
	Joukar, et al.	2017	Neurodegeneration.
	Kaufman, et al.	2015	↑↑ right amygdala volume.
	Piacentino, et al.	2015	↓ memory, ↑ aggressiveness, ↓ anxiety ↑ depression.
	Gomes, et al.	2014	Neurodegeneration.
	Basile, et al.	2013	Neurodegeneration.
	Kanayama, et al.	2013	Cognitive dysfunction.
	Elfverson, et al.	2011	↑ aggressiveness.
	Melloni, Jr., et al.	2010	Anxiety-like behavior.
	Kouvelas, et al.	2008	↓ memory, ↓ anxiety.
	Sato, et al.	2008	↑ aggressiveness.
	Rashid, et al.	2007	↓ memory, ↑ aggressiveness, ↓ anxiety ↑ depression.
	Henderson, et al.	2006	behavioral effects.
Cardiovascular system	Marocolo, et al.	2018	Arrhythmic events; cardiac hypertrophy.
	Rasmussen, et al.	2018	Hypertension; left ventricular hypertrophy.
	Baggish, et al.	2017	↑ left ventricular mass index; ↑ left ventricular walls; myocardial mechanical dysfunction. myocardial infarction.
	Seara, et al.	2017	↑ left ventricular mass index; ↑ left ventricular walls; myocardial mechanical dysfunction.

Table 1. *Cont.*

Organ/System	Author(s)	Year of Publication	Adverse Effects
	Christou, et al.	2016	Abnormal plasma lipoproteins.
	Esperón, et al.	2016	Atherosclerotic plaque.
	Vasilaki, et al.	2016	Cardiac injury
	Akçakoyun et al.	2014	Arrhythmic events.
	Angell, et at.	2012	Arrhythmic events; impairment of cardiac repolarization.
	Golestani, et al.	2012	↓ plasma levels of HDL cholesterol; ↑ LDL cholesterol levels; arrhythmic events.
	Chrostowski, et al.	2011	Increased synthesis of heart muscle, left ventricular hypertrophy and hypertension.
	Riezzo, et al.	2011	Hypertension; left ventricular hypertrophy; pro-atherogenic effects; thrombosis.
	Achar, et al.	2010	↓ plasma levels of HDL cholesterol; ↑ LDL cholesterol levels.
	Alves, et al.	2010	Increased sympathetic activity.
	D'Andrea, et al.	2007	Left ventricular dysfunction.
	Phillis, et al.	2007	Arrhythmic events.
	Chaves, et al.	2006	Oxidative stress.
	Nottin, et al.	2006	↓ left ventricle relaxation properties.
	Pereira-Junior, et al.	2006	Impairment of parasympathetic cardiac modulation; heart rate variability.
Liver	Solimini, et al.	2017	Mitochondrial degeneration of hepatic cells.
	Bond, et al.	2016	Hepatotoxicity.
	Neri, et al.	2011	Hepatotoxicity.
	Schwingel, et al.	2011	Liver fibrosis process.
	Vieira, et al.	2008	Liver fibrosis process.
Urinary system	Brasil, et al.	2015	Kidney fibrosis.
	Riezzo, et al.	2014	Oxidative stress; Accumulation of mesangial matrix.
	D'Errico, et al.	2011	Accumulation of mesangial matrix.
Muscoloskeletal system	Carson, et al.	2015	↑ Muscle mass.
	Marqueti, et al.	2011	Tendon damage.
	Paschoal, et al.	2009	Skeletal muscles injuries.
Reproductive system	Armstrong, et al.	2018	Sexual dysfunctions.
	Barone, et al.	2017	Spermatogenesis alteration.
	El Osta, et al.	2016	Infertility; ↓ number of spermatidis.
	García-Manso, et al.	2016	Changes in number, diameter and thickness of seminiferous tubules.
	Pomara, et al.	2015	Impairment of testosterone production.
	Rahnema, et al.	2014	Hypogonadism.
	Coward, et al.	2013	Hypogonadism.
Hematologic consequences	Chang, et al.	2018	↑ Thrombosis.
	Casavant, et al.	2007	↑ hematocrit, ↑ erythrocytosis.

Abbreviations: ↑ increase; ↓ decrease.

5. Conclusions

This review suggests that AAS misuse and abuse lead to adverse effects in all body tissues and organs. Oxidative stress, apoptosis, and protein synthesis alteration are common mechanisms involved in AAS-related damage in the whole body. This review shows that long-term administration of high doses of AASs may lead to serious consequences, such as hypogonadism, cardiac impairment, neurodegeneration, coronary artery disease and sudden cardiac death. The most reported long-term side effects affect the cardiovascular system, such as cardiomyopathy and atherosclerotic disease. Hypogonadism is a frequent finding in AAS abusers and need to be taken into consideration when AAS use is suspected in order to undertake aggressive treatment [8,120].

Several experimental studies focused on the mechanisms involved in neuropsychiatric effects of AASs. The pathways and the molecular processes are still unclear and need to be clarified [121–124]. In this regard, further studies are needed to assess the epidemiology of antisocial behavior related to AAS assumption and the relationship with other drug consumption.

Moreover, considering that most of the customers are young sportsman and that most of these drugs are easily obtained online, AAS abuse is a considerable public health issue [3].

Clinicians and family doctors should be aware of AAS adverse effects, in order to investigate AAS use in high risk patients, especially in young athletes [121]. In this regard, cardiac imaging may be a helpful tool to assess the presence of subclinical morphological cardiac alterations in AAS abusers. In addition, recent studies reported that miRNAs may play a role in multiple human diseases including AAS adverse effects, suggesting a possible role of these markers in identifying serum or tissue biomarkers with anti-doping potential. However, further studies are needed in this field, given that there is no reliable test to diagnose AAS abuse.

Given the high mortality of atherosclerotic disease and AAS-induced cardiomyopathy, as well as the risk of sudden cardiac death reported in the literature, primary and secondary prevention are crucial in AAS users in order to avoid serious consequences. The scientific community should intensify its efforts to assess the pathophysiology of behavior and cognitive impairment due to long term AAS exposure. Moreover, evidence is urgently required to support the development of a reliable diagnostic tool to identify precociously AAS abuse as well as evidence-based therapy [57,125–131].

Information and education are fundamental tools for AAS misuse preventions. As long as anabolic steroid misuse is popular among young athletes, information campaigns regarding AASs and other doping agents should be encouraged in high schools. In this regard, to prevent the use of AASs public health measures in all settings are crucial. These measures consist of improved knowledge among healthcare workers, proper doping screening tests, educational interventions, and updated legislation.

Although the use of AASs appears to increase the risk of premature death in various categories of patients, further research about this problem is urgently needed [132–139].

Author Contributions: Conceptualization, G.D.A., F.A., G.C., M.S. and A.M.; methodology, G.L.R. and F.M.; software, M.E. and A.L.; validation, G.D.A., M.S. and A.M.; formal analysis, F.A. and G.C.; investigation, N.D.N. and G.L.R.; resources, N.D.N.; data curation, M.S. and F.A.; writing—original draft preparation, G.D.A., F.A. and G.C.; writing—review and editing, M.S. and A.M.; visualization, A.M. and F.M.; supervision, A.L. and N.D.N.; All authors have read and agreed to the published version of the manuscript.

Funding: This research did not receive any specific grant from funding agencies in the public, commercial, or not-for-profit sectors.

Institutional Review Board Statement: Ethical review and approval were waived for this study, because it is a revision of previous published data.

Informed Consent Statement: Not applicable.

Data Availability Statement: Data sharing not applicable, no new data were created or analyzed in this study.

Acknowledgments: The authors thank the Scientific Bureau of the University of Catania for language support.

Conflicts of Interest: The authors declare no conflict of interest.

References

1. Roman, M.; Roman, D.L.; Ostafe, V.; Ciorsac, A.; Isvoran, A. Computational assessment of pharmacokinetics and biological effects of some anabolic and androgen steroids. *Pharm. Res.* **2018**, *35*, 41. [CrossRef] [PubMed]
2. Patanè, F.G.; Liberto, A.; Maglitto, A.N.M.; Malandrino, P.; Esposito, M.; Amico, F.; Cocimano, G.; Li Rosi, G.; Condorelli, D.; Di Nunno, N.; et al. Nandrolone Decanoate: Use, Abuse and Side Effects. *Medicina* **2020**, *56*, 606. [CrossRef] [PubMed]

3. Mullen, C.; Whalley, B.J.; Schifano, F.; Baker, J.S. Anabolic androgenic steroid abuse in the United Kingdom: An update. *Br. J. Pharmacol.* **2020**, *177*, 2180–2198. [CrossRef]
4. Basaria, S. Androgen abuse in athletes: Detection and consequences. *J. Clin. Endocrinol. Metab.* **2010**, *95*, 1533–1543. [CrossRef] [PubMed]
5. El Osta, R.; Almont, T.; Diligent, C.; Hubert, N.; Eschwège, P.; Hubert, J. Anabolic steroids abuse and male infertility. *Basic Clin. Androl.* **2016**, *26*, 2. [CrossRef]
6. Joseph, J.F.; Parr, M.K. Synthetic androgens as designer supplements. *Curr. Neuropharmacol.* **2015**, *13*, 89–100. [CrossRef]
7. Hakansson, A.; Mickelsson, K.; Wallin, C.; Berglund, M. Anabolic androgenic steroids in the general population: User characteristics and associations with substance use. *Eur. Addict. Res.* **2012**, *18*, 83–90. [CrossRef]
8. Kanayama, G.; Kaufman, M.J.; Pope, H.G., Jr. Public health impact of androgens. *Curr. Opin. Endocrinol. Diabetes Obes.* **2018**, *25*, 218. [CrossRef]
9. Rahnema, C.D.; Crosnoe, L.E.; Kim, E.D. Designer steroids-over-the-counter supplements and their androgenic component: Review of an increasing problem. *Andrology* **2015**, *3*, 150–155. [CrossRef]
10. Kanayama, G.; Hudson, J.I.; Pope, H.G., Jr. Illicit anabolic–androgenic steroid use. *Horm. Behav.* **2010**, *58*, 111–121. [CrossRef]
11. Anawalt, B.D. Diagnosis and management of anabolic androgenic steroid use. *J. Clin. Endocrinol. Metab.* **2019**, *104*, 2490–2500. [CrossRef] [PubMed]
12. Kicman, A.T. Pharmacology of anabolic steroids. *Br. J. Pharmacol.* **2008**, *154*, 502–521. [CrossRef]
13. Sessa, F.; Salerno, M.; Di Mizio, G.; Bertozzi, G.; Messina, G.; Tomaiuolo, B.; Pisanelli, D.; Maglietta, F.; Ricci, P.; Pomara, C. Anabolic androgenic steroids: Searching new molecular biomarkers. *Front. Pharmacol.* **2018**, *9*, 1321. [CrossRef] [PubMed]
14. Alsiö, J.; Birgner, C.; Björkblom, L.; Isaksson, P.; Bergström, L.; Schiöth, H.B.; Lindblom, J. Impact of nandrolone decanoate on gene expression in endocrine systems related to the adverse effects of anabolic androgenic steroids. *Basic Clin. Pharmacol. Toxicol.* **2009**, *105*, 307–314. [CrossRef] [PubMed]
15. Cheung, A.S.; Grossmann, M. Physiological basis behind ergogenic effects of anabolic androgens. *Mol. Cell. Endocrinol.* **2018**, *464*, 14–20. [CrossRef]
16. Fortunato, R.S.; Marassi, M.P.; Chaves, E.A.; Nascimento, J.H.; Rosenthal, D.; Carvalho, D.P. Chronic administration of anabolic androgenic steroid alters murine thyroid function. *Med. Sci. Sports Exerc.* **2006**, *38*, 256–261. [CrossRef] [PubMed]
17. Nieschlag, E.; Vorona, E. Doping with anabolic androgenic steroids(AAS): Adverse effects on non-reproductive organs and functions. *Rev. Endocr. Metab. Disord.* **2015**, *16*, 199–211. [CrossRef]
18. Arazi, H.; Mohammadjafari, H.; Asadi, A. Use of anabolic androgenic steroids produces greater oxidative stress responses to resistance exercise in strength-trained men. *Toxicol. Rep.* **2017**, *4*, 282–286. [CrossRef]
19. Frankenfeld, S.P.; Oliveira, L.P.; Ortenzi, V.H.; Rego-Monteiro, I.C.; Chaves, E.A.; Ferreira, A.C.; Leitao, A.C.; Carvalho, D.P.; Fortunato, R.S. The anabolic androgenic steroid nandrolone decanoate disrupts redox homeostasis in liver, heart and kidney of male Wistar rats. *PLoS ONE* **2014**, *9*, e102699. [CrossRef]
20. Frankenfeld, S.P.; de Oliveira, L.P.; Ignacio, D.L.; Coelho, R.G.; Mattos, M.N.; Ferreira, A.C.; Carvalho, D.P.; Fortunato, R.S. Nandrolone decanoate inhibits gluconeogenesis and decreases fasting glucose in Wistar male rats. *J. Endocrinol.* **2014**, *220*, 143–153. [CrossRef]
21. Chaves, E.A.; Pereira, P.P., Jr.; Fortunato, R.S.; Masuda, M.O.; de Carvalho, A.C.; de Carvalho, D.P.; Oliveira, M.F.; Nascimento, J.H. Nandrolone decanoate impairs exercise-induced cardioprotection: Role of antioxidant enzymes. *J. Steroid Biochem. Mol. Biol.* **2006**, *99*, 223–230. [CrossRef] [PubMed]
22. Cerretani, D.; Neri, M.; Cantatore, S.; Ciallella, C.; Riezzo, I.; Turillazzi, E.; Fineschi, V. Looking for organ damages due to anabolic-androgenic steroids(AAS): Is oxidative stress the culprit? *Mini Rev. Org. Chem.* **2013**, *10*, 393–399. [CrossRef]
23. Turillazzi, E.; Neri, M.; Cerretani, D.; Cantatore, S.; Frati, P.; Moltoni, L.; Busardò, F.P.; Pomara, C.; Riezzo, I.; Fineschi, V. Lipid peroxidation and apoptotic response in rat brain areas induced by long-term administration of nandrolone: The mutual crosstalk between ROS and NF-kB. *J. Cell. Mol. Med.* **2016**, *20*, 601–612. [CrossRef] [PubMed]
24. Basile, J.R.; Binmadi, N.O.; Zhou, H.; Yang, Y.; Paoli, A.; Proia, P. Supraphysiological doses of performance enhancing anabolic-androgenic steroids exert direct toxic effects on neuron-like cells. *Front. Cell Neurosci.* **2013**, *7*, 69. [CrossRef]
25. Joukar, S.; Vahidi, R.; Farsinejad, A.; Asadi-Shekaari, M.; Shahouzehi, B. Ameliorative effects of endurance exercise with two different intensities on nandrolone decanoate-induced neurodegeneration in rats: Involving redox and apoptotic systems. *Neurotox. Res.* **2017**, *32*, 41–49. [CrossRef]
26. D'Errico, S.; Di Battista, B.; Di Paolo, M.; Fiore, C.; Pomara, C. Renal heat shock proteins over-expression due to anabolic androgenic steroids abuse. *Mini Rev. Med. Chem.* **2011**, *11*, 446–450. [CrossRef]
27. Chrostowski, K.; Kwiatkowska, D.; Pokrywka, A.; Stanczyk, D.; Wójcikowska-Wójcik, B.; Grucza, R. Renin-angiotensin-aldosterone system in bodybuilders using supraphysiological doses of anabolic-androgenic steroids. *Biol. Sport* **2011**, *28*, 11. [CrossRef]
28. D'Andrea, A.; Caso, P.; Salerno, G.; Scarafile, R.; De Corato, G.; Mita, C.; Di Salvo, G.; Severino, S.; Cuomo, S.; Liccardo, B.; et al. Left ventricular early myocardial dysfunction after chronic misuse of anabolic androgenic steroids: A Doppler myocardial and strain imaging analysis. *Br. J. Sports Med.* **2007**, *41*, 149–155. [CrossRef]

29. Lopes, R.A.; Neves, K.B.; Pestana, C.R.; Queiroz, A.L.; Zanotto, C.Z.; Chignalia, A.Z.; Valim, Y.M.; Silveira, L.R.; Curti, C.; Tostes, R.C. Testosterone induces apoptosis in vascular smooth muscle cells via extrinsic apoptotic pathway with mitochondria-generated reactive oxygen species involvement. *Am. J. Physiol. Heart Circ. Physiol.* **2014**, *306*, H1485–H1494. [CrossRef]
30. Hackett, D.A.; Johnson, N.A.; Chow, C.M. Training practices and ergogenic aids used by male bodybuilders. *J. Strength Cond. Res.* **2013**, *27*, 1609–1617. [CrossRef]
31. Sjöqvist, F.; Garle, M.; Rane, A. Use of doping agents, particularly anabolic steroids, in sports and society. *Lancet* **2008**, *371*, 1872–1882. [CrossRef]
32. Turillazzi, E.; Perilli, G.; Di Paolo, M.; Neri, M.; Riezzo, I.; Fineschi, V. Side effects of AAS abuse: An overview. *Mini Rev. Med. Chem.* **2011**, *11*, 374–389. [CrossRef] [PubMed]
33. Van Amsterdam, J.; Opperhuizen, A.; Hartgens, F. Adverse health effects of anabolic–androgenic steroids. *Regul. Toxicol. Pharmacol.* **2010**, *57*, 117–123. [CrossRef] [PubMed]
34. Petersson, A.; Garle, M.; Holmgren, P.; Druid, H.; Krantz, P.; Thiblin, I. Toxicological findings and manner of death in autopsied users of anabolic androgenic steroids. *Drug Alcohol Depend.* **2006**, *81*, 241–249. [CrossRef] [PubMed]
35. Frati, P.; Busardo, F.P.; Cipolloni, L.; De Dominicis, E.; Fineschi, V. Anabolic androgenic steroid(AAS) related deaths: Autoptic, histopathological and toxicological findings. *Curr. Neuropharmacol.* **2015**, *13*, 146–159. [CrossRef]
36. Darke, S.; Torok, M.; Duflou, J. Sudden or unnatural deaths involving anabolic-androgenic steroids. *J. Forensic Sci.* **2014**, *59*, 1025–1028. [CrossRef]
37. Di Paolo, M.; Agozzino, M.; Toni, C.; Luciani, A.B.; Molendini, L.; Scaglione, M.; Inzani, F.; Pasotti, M.; Buffi, F.; Arbustini, E. Sudden anabolic steroid abuse-related death in athletes. *Int. J. Cardiol.* **2007**, *114*, 114–117. [CrossRef] [PubMed]
38. Hernández-Guerra, A.I.; Tapia, J.; Menéndez-Quintanal, L.M.; Lucena, J.S. Sudden cardiac death in anabolic androgenic steroids abuse: Case report and literature review. *Forensic Sci. Res.* **2019**, *4*, 267–273. [CrossRef]
39. Caraci, F.; Pistarà, V.; Corsaro, A.; Tomasello, F.; Giuffrida, M.L.; Sortino, M.A.; Nicoletti, F.; Copani, A. Neurotoxic properties of the anabolic androgenic steroids nandrolone and methandrostenolone in primary neuronal cultures. *J. Neurosci. Res.* **2011**, *89*, 592–600. [CrossRef]
40. Karimooy, F.N.; Bideskan, A.E.; Pour, A.M.; Hoseini, S.M. Neurotoxic Effects of Stanozolol on Male Rats' Hippocampi: Does Stanozolol cause apoptosis? *Biomol. Concepts* **2019**, *10*, 73–81. [CrossRef]
41. Gomes, F.G.; Fernandes, J.; Campos, D.V.; Cassilhas, R.C.; Viana, G.M.; D'Almeida, V.; de Moraes Rego, M.K.; Buainain, P.I.; Cavalheiro, E.A.; Arida, R.M. The beneficial effects of strength exercise on hippocampal cell proliferation and apoptotic signaling is impaired by anabolic androgenic steroids. *Psychoneuroendocrinology* **2014**, *50*, 106–117. [CrossRef] [PubMed]
42. Sessa, F.; Salerno, M.; Cipolloni, L.; Bertozzi, G.; Messina, G.; Di Mizio, G.; Asmundo, A.; Pomara, C. Anabolic-androgenic steroids and brain injury: miRNA evaluation in users compared to cocaine abusers and elderly people. *Aging* **2020**, *12*. [CrossRef]
43. Bjørnebekk, A.; Westlye, L.T.; Walhovd, K.B.; Jørstad, M.L.; Sundseth, Ø.Ø.; Fjell, A.M. Cognitive performance and structural brain correlates in long-term anabolic-androgenic steroid exposed and nonexposed weightlifters. *Neuropsychology* **2019**, *33*, 547. [CrossRef] [PubMed]
44. Bjørnebekk, A.; Walhovd, K.B.; Jørstad, M.L.; Due-Tønnessen, P.; Hullstein, I.R.; Fjell, A.M. Structural brain imaging of long-term anabolic-androgenic steroid users and nonusing weightlifters. *Biol. Psychiatry* **2017**, *82*, 294–302. [CrossRef]
45. Kaufman, M.J.; Janes, A.C.; Hudson, J.I.; Brennan, B.P.; Kanayama, G.; Kerrigan, A.R.; Jensen, J.E.; Pope, H.G., Jr. Brain and cognition abnormalities in long-term anabolic-androgenic steroid users. *Drug Alcohol Depend.* **2015**, *152*, 47–56. [CrossRef]
46. Kanayama, G.; Kean, J.; Hudson, J.I.; Pope, H.G., Jr. Cognitive deficits in long-term anabolic-androgenic steroid users. *Drug Alcohol Depend.* **2013**, *130*, 208–214. [CrossRef]
47. Kaufman, M.J.; Kanayama, G.; Hudson, J.I.; Pope, H.G., Jr. Supraphysiologic-dose anabolic–androgenic steroid use: A risk factor for dementia? *Neurosci. Biobehav. Rev.* **2019**, *100*, 180–207. [CrossRef]
48. Piacentino, D.; Kotzalidis, G.D.; Del Casale, A.; Aromatario, M.R.; Pomara, C.; Girardi, P.; Sani, G. Anabolic-androgenic steroid use and psychopathology in athletes. A systematic review. *Curr. Neuropharmacol.* **2015**, *13*, 101–121. [CrossRef]
49. Rashid, H.; Ormerod, S.; Day, E. Anabolic androgenic steroids: What the psychiatrist needs to know. *Adv. Psychiatr. Treat.* **2007**, *13*, 203–211. [CrossRef]
50. Joksimovic, J.; Selakovic, D.; Jovicic, N.; Mitrovic, S.; Mihailovic, V.; Katanic, J.; Milovanovic, D.; Rosic, G. Exercise Attenuates Anabolic Steroids-Induced Anxiety via Hippocampal NPY and MC4 Receptor in Rats. *Front. Neurosci.* **2019**, *13*, 172. [CrossRef]
51. Hauger, L.E.; Sagoe, D.; Vaskinn, A.; Arnevik, E.A.; Leknes, S.; Jørstad, M.L.; Bjørnebekk, A. Anabolic androgenic steroid dependence is associated with impaired emotion recognition. *Psychopharmacology* **2019**, *236*, 2667–2676. [CrossRef] [PubMed]
52. Almeida, O.P.; Yeap, B.B.; Hankey, G.J.; Jamrozik, K.; Flicker, L. *Low Free Testosterone Concentration as a Potentially Treatable Cause of Depressive Symptoms in Older Men*; Arch. Gen. Psychiatry: Chicago, IL, USA, 2008; Volume 12, pp. 283–289. ISSN 0003-990X.
53. Bueno, A.; Carvalho, F.B.; Gutierres, J.M.; Lhamas, C.; Andrade, C.M. A comparative study of the effect of the dose and exposure duration of anabolic androgenic steroids on behavior, cholinergic regulation, and oxidative stress in rats. *PLoS ONE* **2017**, *12*, e0177623. [CrossRef] [PubMed]
54. Henderson, L.P.; Penatti, C.A.; Jones, B.L.; Yang, P.; Clark, A.S. Anabolic androgenic steroids and forebrain GABAergic transmission. *Neuroscience* **2006**, *138*, 793–799. [CrossRef] [PubMed]
55. Kouvelas, D.; Pourzitaki, C.; Papazisis, G.; Dagklis, T.; Dimou, K.; Kraus, M.M. Nandrolone abuse decreases anxiety and impairs memory in rats via central androgenic receptors. *Int. J. Neuropsychopharmacol.* **2008**, *11*, 925–934. [CrossRef] [PubMed]

56. Melloni, R.H., Jr.; Ricci, L.A. Adolescent exposure to anabolic/androgenic steroids and the neurobiology of offensive aggression: A hypothalamic neural model based on findings in pubertal Syrian hamsters. *Horm. Behav.* **2010**, *58*, 177–191. [CrossRef] [PubMed]
57. Quaglio, G.; Fornasiero, A.; Mezzelani, P.; Moreschini, S.; Lugoboni, F.; Lechi, A. Anabolic steroids: Dependence and complications of chronic use. *Intern. Emerg. Med.* **2009**, *4*, 289–296. [CrossRef]
58. Elfverson, M.; Johansson, T.; Zhou, Q.; Le Grevès, P.; Nyberg, F. Chronic administration of the anabolic androgenic steroid nandrolone alters neurosteroid action at the sigma-1 receptor but not at the sigma-2 or NMDA receptors. *Neuropharmacology* **2011**, *61*, 1172–1181. [CrossRef]
59. Sato, S.M.; Schulz, K.M.; Sisk, C.L.; Wood, R.I. Adolescents and androgens, receptors and rewards. *Horm. Behav.* **2008**, *53*, 647–658. [CrossRef]
60. Bertozzi, G.; Sessa, F.; Albano, G.D.; Sani, G.; Maglietta, F.; Roshan, M.H.; Li Volti, G.; Bernardini, R.; Avola, R.; Pomara, C.; et al. The role of anabolic androgenic steroids in disruption of the physiological function in discrete areas of the central nervous system. *Mol. Neurobiol.* **2018**, *55*, 5548–5556. [CrossRef]
61. Bertozzi, G.; Salerno, M.; Pomara, C.; Sessa, F. Neuropsychiatric and Behavioral Involvement in AAS Abusers. A Literature Review. *Medicina* **2019**, *55*, 396. [CrossRef]
62. Liu, J.D.; Wu, Y.Q. Anabolic-androgenic steroids and cardiovascular risk. *Chin. Med. J.* **2019**, *132*, 2229. [CrossRef] [PubMed]
63. Vasilaki, F.; Tsitsimpikou, C.; Tsarouhas, K.; Germanakis, I.; Tzardi, M.; Kavvalakis, M.; Ozcagli, E.; Kouretas, D.; Tsatsakis, A.M. Cardiotoxicity in rabbits after long-term nandrolone decanoate administration. *Toxicol. Lett.* **2016**, *241*, 143–151. [CrossRef] [PubMed]
64. Pereira-Junior, P.P.; Chaves, E.A.; Costa-e-Sousa, R.H.; Masuda, M.O.; de Carvalho, A.C.; Nascimento, J.H. Cardiac autonomic dysfunction in rats chronically treated with anabolic steroid. *Eur. J. Appl. Physiol.* **2006**, *96*, 487–494. [CrossRef] [PubMed]
65. Riezzo, I.; Di Paolo, M.; Neri, M.; Bello, S.; Cantatore, S.; D'Errico, S.; Dinucci, D.; Parente, R.; Pomara, C.; Rabozzi, R.; et al. Anabolic steroid-and exercise-induced cardio-depressant cytokines and myocardial β1 receptor expression in CD1 mice. *Curr. Pharm. Biotechnol.* **2011**, *12*, 275–284. [CrossRef]
66. Christou, G.A.; Christou, K.A.; Nikas, D.N.; Goudevenos, J.A. Acute myocardial infarction in a young bodybuilder taking anabolic androgenic steroids: A case report and critical review of the literature. *Eur. J. Prev. Cardiol.* **2016**, *23*, 1785–1796. [CrossRef]
67. Achar, S.; Rostamian, A.; Narayan, S.M. Cardiac and metabolic effects of anabolic-androgenic steroid abuse on lipids, blood pressure, left ventricular dimensions, and rhythm. *Am. J. Cardiol.* **2010**, *106*, 893–901. [CrossRef]
68. Golestani, R.; Slart, R.H.; Dullaart, R.P.; Glaudemans, A.W.; Zeebregts, C.J.; Boersma, H.H.; Tio, R.A.; Dierckx, R.A. Adverse cardiovascular effects of anabolic steroids: Pathophysiology imaging. *Eur. J. Clin. Investig.* **2012**, *42*, 795. [CrossRef]
69. Alves, M.J.; Dos Santos, M.R.; Dias, R.G.; Akiho, C.A.; Laterza, M.C.; Rondon, M.U.; Moreau, R.L.; Negrao, C.E. Abnormal neurovascular control in anabolic androgenic steroids users. *Med. Sci. Sports Exerc.* **2010**, *42*, 865–871. [CrossRef]
70. Baggish, A.L.; Weiner, R.B.; Kanayama, G.; Hudson, J.I.; Picard, M.H.; Hutter, A.M., Jr.; Pope, H.G., Jr. Long-term anabolic-androgenic steroid use is associated with left ventricular dysfunction. *Circ. Heart Fail.* **2010**, *3*, 472–476. [CrossRef]
71. Seara, F.A.; Barbosa, R.A.; de Oliveira, D.F.; Gran da Silva, D.L.; Carvalho, A.B.; Freitas Ferreira, A.C.; Nascimento, J.H.; Olivares, E.L. Administration of anabolic steroid during adolescence induces long-term cardiac hypertrophy and increases susceptibility to ischemia/reperfusion injury in adult Wistar rats. *J. Steroid Biochem. Mol. Biol.* **2017**, *171*, 34–42. [CrossRef]
72. Rasmussen, J.J.; Schou, M.; Madsen, P.L.; Selmer, C.; Johansen, M.L.; Ulriksen, P.S.; Dreyer, T.; Kumler, T.; Plesner, L.L.; Faber, J.; et al. Cardiac systolic dysfunction in past illicit users of anabolic androgenic steroids. *Am. Heart J.* **2018**, *203*, 49–56. [CrossRef]
73. Nottin, S.; Nguyen, L.D.; Terbah, M.; Obert, P. Cardiovascular effects of androgenic anabolic steroids in male bodybuilders determined by tissue Doppler imaging. *Am. J. Cardiol.* **2006**, *97*, 912–915. [CrossRef]
74. Rocha, F.L.; Carmo, E.C.; Roque, F.R.; Hashimoto, N.Y.; Rossoni, L.V.; Frimm, C.; Aneas, I.; Negrao, C.E.; Krieger, J.E.; Oliveira, E.M. Anabolic steroids induce cardiac renin-angiotensin system and impair the beneficial effects of aerobic training in rats. *Am. J. Physiol. Heart Circ. Physiol.* **2007**, *293*, H3575–H3583. [CrossRef]
75. Akçakoyun, M.; Alizade, E.; Gündoğdu, R.; Bulut, M.; Tabakci, M.M.; Açar, G.; Avci, A.; Simsek, Z.; Fidan, S.; Demir, S.; et al. Long-term anabolic androgenic steroid use is associated with increased atrial electromechanical delay in male bodybuilders. *Biomed. Res. Int.* **2014**. [CrossRef]
76. Angell, P.; Chester, N.; Green, D.; Somauroo, J.; Whyte, G.; George, K. Anabolic steroids and cardiovascular risk. *Sports Med.* **2012**, *42*, 119–134. [CrossRef]
77. Marocolo, M.; Silva-Neto, J.A.; Neto, O.B. Acute interruption of treatment with nandrolone decanoate is not sufficient to reverse cardiac autonomic dysfunction and ventricular repolarization disturbances in rats. *Steroids* **2018**, *132*, 12–17. [CrossRef]
78. Phillis, B.D.; Abeywardena, M.Y.; Adams, M.J.; Kennedy, J.A.; Irvine, R.J. Nandrolone potentiates arrhythmogenic effects of cardiac ischemia in the rat. *Toxicol. Sci.* **2007**, *99*, 605–611. [CrossRef]
79. Olivares, E.L.; Silveira, A.L.; Fonseca, F.V.; Silva-Almeida, C.; Côrtes, R.S.; Pereira-Junior, P.P.; Nascimento, J.H.; Reis, L.C. Administration of an anabolic steroid during the adolescent phase changes the behavior, cardiac autonomic balance and fluid intake in male adult rats. *Physiol. Behav.* **2014**, *126*, 15–24. [CrossRef]
80. Esperón, C.G.; Lopez-Cancio, E.; Bermejo, P.G.; Dávalos, A. Anabolic Androgenic Steroids and Stroke. In *Neuropathology of Drug Addictions and Substance Misuse*, 2nd ed.; Preedy, V.R., Ed.; Academic Press: London, UK, 2016; pp. 981–990. ISBN 978-0-12-800212-4.

81. Baggish, A.L.; Weiner, R.B.; Kanayama, G.; Hudson, J.I.; Lu, M.T.; Hoffmann, U.; Pope, H.G., Jr. Cardiovascular toxicity of illicit anabolic-androgenic steroid use. *Circulation* **2017**, *135*, 1991–2002. [CrossRef]
82. Bond, P.; Llewellyn, W.; Van Mol, P. Anabolic androgenic steroid-induced hepatotoxicity. *Med. Hypotheses* **2016**, *93*, 150–153. [CrossRef]
83. Neri, M.; Bello, S.; Bonsignore, A.; Cantatore, S.; Riezzo, I.; Turillazzi, E.; Fineschi, V. Anabolic androgenic steroids abuse and liver toxicity. *Mini Rev. Med. Chem.* **2011**, *11*, 430–437. [CrossRef] [PubMed]
84. Solimini, R.; Rotolo, M.C.; Mastrobattista, L.; Mortali, C.; Minutillo, A.; Pichini, S.; Pacifici, I.; Palmi, I. Hepatotoxicity associated with illicit use of anabolic androgenic steroids in doping. *Eur. Rev. Med. Pharmacol. Sci.* **2017**, *21*, 7–16. [PubMed]
85. Vieira, R.P.; Franca, R.F.; Damaceno-Rodrigues, N.R.; Dolhnikoff, M.; Caldini, E.G.; Carvalho, C.R.; Ribeiro, W. Dose-dependent hepatic response to subchronic administration of nandrolone decanoate. *Med. Sci. Sports Exerc.* **2008**, *40*, 842. [CrossRef] [PubMed]
86. Schwingel, P.A.; Cotrim, H.P.; Salles, B.R.; Almeida, C.E.; dos Santos, C.R., Jr.; Nachef, B.; Andrade, A.R.; Zoppi, C.C. Anabolic-androgenic steroids: A possible new risk factor of toxicant-associated fatty liver disease. *Liver Int.* **2011**, *31*, 348–353. [CrossRef] [PubMed]
87. Riezzo, I.; Turillazzi, E.; Bello, S.; Cantatore, S.; Cerretani, D.; Di Paolo, M.; Fiaschi, A.I.; Frati, P.; Neri, M.; Pedretti, M.; et al. Chronic nandrolone administration promotes oxidative stress, induction of pro-inflammatory cytokine and TNF-α mediated apoptosis in the kidneys of CD1 treated mice. *Toxicol. Appl. Pharmacol.* **2014**, *280*, 97–106. [CrossRef]
88. Kahal, A.; Allem, R. Reversible effects of anabolic steroid abuse on cyto-architectures of the heart, kidneys and testis in adult male mice. *Biomed. Pharmacother.* **2018**, *106*, 917–922. [CrossRef]
89. Brasil, G.A.; de Lima, E.M.; do Nascimento, A.M.; Caliman, I.F.; de Medeiros, A.R.; Silva, M.S.; de Abreu, G.R.; dos Reis, A.M.; de Andrade, T.U.; Bissoli, N.S. Nandrolone decanoate induces cardiac and renal remodeling in female rats, without modification in physiological parameters: The role of ANP system. *Life Sci.* **2015**, *137*, 65–73. [CrossRef]
90. Josiak, K.; Jankowska, E.A.; Piepoli, M.F.; Banasiak, W.; Ponikowski, P. Skeletal myopathy in patients with chronic heart failure: Significance of anabolic-androgenic hormones. *J. Cachexia Sarcopenia Muscle* **2014**, *5*, 287–296. [CrossRef]
91. Hoseini, L.; Roozbeh, J.; Sagheb, M.; Karbalay-Doust, S.; Noorafshan, A. Nandrolone decanoate increases the volume but not the length of the proximal and distal convoluted tubules of the mouse kidney. *Micron* **2009**, *40*, 226–230. [CrossRef]
92. Carson, J.A.; Manolagas, S.C. Effects of sex steroids on bones and muscles: Similarities, parallels, and putative interactions in health and disease. *Bone* **2015**, *80*, 67–78. [CrossRef]
93. Casavant, M.J.; Blake, K.; Griffith, J.; Yates, A.; Copley, L.M. Consequences of use of anabolic androgenic steroids. *Pediatr. Clin. N. Am.* **2007**, *54*, 677–690. [CrossRef]
94. Andrews, M.A.; Magee, C.D.; Combest, T.M.; Allard, R.J.; Douglas, K.M. Physical effects of anabolic-androgenic steroids in healthy exercising adults: A systematic review and meta-analysis. *Curr. Sports Med. Rep.* **2018**, *17*, 232–241. [CrossRef]
95. Caminiti, G.; Volterrani, M.; Iellamo, F.; Marazzi, G.; Massaro, R.; Miceli, M.; Mammi, C.; Piepoli, M.; Fini, M.; Rosano, G.M. Effect of long-acting testosterone treatment on functional exercise capacity, skeletal muscle performance, insulin resistance, and baroreflex sensitivity in elderly patients with chronic heart failure: A double-blind, placebo-controlled, randomized study. *J. Am. Coll. Cardiol.* **2009**, *54*, 919–927. [CrossRef]
96. Marqueti, R.C.; Prestes, J.; Stotzer, U.S.; Paschoal, M.; Leite, R.D.; Perez, S.E.; de Araujo, H.S. MMP-2, jumping exercise and nandrolone in skeletal muscle. *Int. J. Sports Med.* **2008**, *29*, 559–563. [CrossRef]
97. Paschoal, M.; de Cássia Marqueti, R.; Perez, S.; Selistre-de-Araujo, H.S. Nandrolone inhibits VEGF mRNA in rat muscle. *Int. J. Sports Med.* **2009**, *30*, 775–778. [CrossRef]
98. Marqueti, R.C.; Prestes, J.; Wang, C.C.; Ramos, O.H.; Perez, S.E.; Nakagaki, W.R.; Carvalho, H.F.; Selistre-de-Araujo, H.S. Biomechanical responses of different rat tendons to nandrolone decanoate and load exercise. *Scand. J. Med. Sci. Sports* **2011**, *21*, e91–e99. [CrossRef]
99. Marqueti, R.C.; Prestes, J.; Paschoal, M.; Ramos, O.H.; Perez, S.E.; Carvalho, H.F.; Selistre-de-Araujo, H.S. Matrix metallopeptidase 2 activity in tendon regions: Effects of mechanical loading exercise associated to anabolic-androgenic steroids. *Eur. J. Appl. Physiol.* **2008**, *104*, 1087. [CrossRef]
100. Pomara, C.; Barone, R.; Marino Gammazza, A.; Sangiorgi, C.; Barone, F.; Pitruzzella, A.; Locorotondo, N.; Di Gaudio, F.; Salerno, M.; Maglietta, F.; et al. Effects of nandrolone stimulation on testosterone biosynthesis in leydig cells. *J. Cell. Physiol.* **2016**, *231*, 1385–1391. [CrossRef]
101. Armstrong, J.M.; Avant, R.A.; Charchenko, C.M.; Westerman, M.E.; Ziegelmann, M.J.; Miest, T.S.; Trost, L.W. Impact of anabolic androgenic steroids on sexual function. *Transl. Androl. Urol.* **2018**, *7*, 483. [CrossRef]
102. Coward, R.M.; Rajanahally, S.; Kovac, J.R.; Smith, R.P.; Pastuszak, A.W.; Lipshultz, L.I. Anabolic steroid induced hypogonadism in young men. *J. Urol.* **2013**, *190*, 2200–2205. [CrossRef]
103. Barone, R.; Pitruzzella, A.; Marino Gammazza, A.; Rappa, F.; Salerno, M.; Barone, F.; Sangiorgi, C.; D'Amico, D.; Locorotondo, N.; Di Gaudio, F.; et al. Nandrolone decanoate interferes with testosterone biosynthesis altering blood–testis barrier components. *J. Cell. Mol. Med.* **2017**, *21*, 1636–1647. [CrossRef] [PubMed]
104. García-Manso, J.M.; Esteve, T.V. Consequences of the Use of Anabolic-Androgenic Steroids for Male Athletes' Fertility. In *Exercise and Human Reproduction*; Vaamonde, D., du Plessis, S.S., Agarwal, A., Eds.; Springer: New York, NY, USA, 2016; pp. 153–165. ISBN 978-1-4939-3402-7.

105. Rahnema, C.D.; Lipshultz, L.I.; Crosnoe, L.E.; Kovac, J.R.; Kim, E.D. Anabolic steroid–induced hypogonadism: Diagnosis and treatment. *Fertil. Steril.* **2014**, *101*, 1271–1279. [CrossRef] [PubMed]
106. Tatem, A.J.; Beilan, J.; Kovac, J.R.; Lipshultz, L.I. Management of anabolic steroid-induced infertility: Novel strategies for fertility maintenance and recovery. *World J. Mens. Health* **2020**, *38*, 141–150. [CrossRef] [PubMed]
107. Agledahl, I.; Brodin, E.; Svartberg, J.; Hansen, J.B. Plasma free tissue factor pathway inhibitor(TFPI) levels and TF-induced thrombin generation ex vivo in men with low testosterone levels. *Thromb. Haemost.* **2009**, *101*, 471–477. [CrossRef]
108. Liljeqvist, S.; Helldén, A.; Bergman, U.; Söderberg, M. Pulmonary embolism associated with the use of anabolic steroids. *Eur. J. Intern. Med.* **2008**, *19*, 214–215. [CrossRef]
109. Thiblin, I.; Garmo, H.; Garle, M.; Holmberg, L.; Byberg, L.; Michaëlsson, K.; Gedeborg, R. Anabolic steroids and cardiovascular risk: A national population-based cohort study. *Drug Alcohol Depend.* **2015**, *152*, 87–92. [CrossRef]
110. Jin, H.; Lin, J.; Fu, L.; Mei, Y.F.; Peng, G.; Tan, X.; Wang, D.M.; Wang, W.; Li, Y.G. Physiological testosterone stimulates tissue plasminogen activator and tissue factor pathway inhibitor and inhibits plasminogen activator inhibitor type 1 release in endothelial cells. *Biochem. Cell Biol.* **2007**, *85*, 246–251. [CrossRef]
111. Chang, S.; Münster, A.B.; Gram, J.; Sidelmann, J.J. Anabolic androgenic steroid abuse: The effects on thrombosis risk, coagulation, and fibrinolysis. *Semin. Thromb. Hemost.* **2018**, *44*, 734–746. [CrossRef]
112. Bertozzi, G.; Sessa, F.; Maglietta, F.; Cipolloni, L.; Salerno, M.; Fiore, C.; Fortarezza, P.; Ricci, P.; Turillazzi, E.; Pomara, C. Immunodeficiency as a side effect of anabolic androgenic steroid abuse: A case of necrotizing myofasciitis. *Forensic Sci. Med. Pathol.* **2019**, *15*, 616–621. [CrossRef]
113. Salerno, M.; Cascio, O.; Bertozzi, G.; Sessa, F.; Messina, A.; Monda, V.; Cipolloni, L.; Biondi, A.; Daniele, A.; Pomara, C. Anabolic androgenic steroids and carcinogenicity focusing on Leydig cell: A literature review. *Oncotarget* **2018**, *9*, 19415. [CrossRef]
114. Hashimoto, H.; Vertino, P.M.; Cheng, X. Molecular coupling of DNA methylation and histone methylation. *Epigenomics* **2010**, *2*, 657–669. [CrossRef] [PubMed]
115. Medei, E.; Marocolo, M.; de Carvalho Rodrigues, D.; Arantes, P.C.; Takiya, C.M.; Silva, J.; Rondinelli, E.; Goldenberg, R.C.; Campos de Carvalho, A.C.; Nascimento, J.H. Chronic treatment with anabolic steroids induces ventricular repolarization disturbances: Cellular, ionic and molecular mechanism. *J. Mol. Cell. Cardiol.* **2010**, *49*, 165–175. [CrossRef] [PubMed]
116. Schwarzenbach, H. Impact of physical activity and doping on epigenetic gene regulation. *Drug Test. Anal.* **2011**, *3*, 682–687. [CrossRef] [PubMed]
117. Souza, L.D.; da Cruz, L.A.; Cerqueira, E.D.; Meireles, J.R. Micronucleus as biomarkers of cancer risk in anabolic androgenic steroids users. *Hum. Exp. Toxicol.* **2017**, *36*, 302–310. [CrossRef] [PubMed]
118. Agriesti, F.; Tataranni, T.; Pacelli, C.; Scrima, R.; Laurenzana, I.; Ruggieri, V.; Cela, O.; Mazzoccoli, C.; Salerno, M.; Sessa, F.; et al. Nandrolone induces a stem cell-like phenotype in human hepatocarcinoma-derived cell line inhibiting mitochondrial respiratory activity. *Sci. Rep.* **2020**, *10*, 1–17. [CrossRef]
119. Tentori, L.; Graziani, G. Doping with growth hormone/IGF-1, anabolic steroids or erythropoietin: Is there a cancer risk? *Pharmacol. Res.* **2007**, *55*, 359–369. [CrossRef]
120. Kanayama, G.; Pope, H.G., Jr.; Hudson, J.I. Associations of anabolic-androgenic steroid use with other behavioral disorders: An analysis using directed acyclic graphs. *Psychol. Med.* **2018**, *48*, 2601–2608. [CrossRef]
121. Brooks, J.H.; Ahmad, I.; Easton, G. Anabolic steroid use. *Br. Med. J.* **2016**, *355*. [CrossRef]
122. Grönbladh, A.; Nylander, E.; Hallberg, M. The neurobiology and addiction potential of anabolic androgenic steroids and the effects of growth hormone. *Brain Res. Bull.* **2016**, *126*, 127–137. [CrossRef]
123. Riezzo, I.; De Carlo, D.; Neri, M.; Nieddu, A.; Turillazzi, E.; Fineschi, V. Heart disease induced by AAS abuse, using experimental mice/rats models and the role of exercise-induced cardiotoxicity. *Mini Rev. Med. Chem.* **2011**, *11*, 409–424. [CrossRef]
124. Pagonis, T.A.; Angelopoulos, N.V.; Koukoulis, G.N.; Hadjichristodoulou, C.S. Psychiatric side effects induced by supraphysiological doses of combinations of anabolic steroids correlate to the severity of abuse. *Eur. Psychiatry* **2006**, *21*, 551–562. [CrossRef]
125. Baker, J.S.; Graham, M.R.; Davies, B. Steroid and prescription medicine abuse in the health and fitness community: A regional study. *Eur. J. Intern. Med.* **2006**, *17*, 479–484. [CrossRef]
126. Nagata, J.M.; Ganson, K.T.; Gorrell, S.; Mitchison, D.; Murray, S.B. Association between legal performance-enhancing substances and use of anabolic-androgenic steroids in young adults. *JAMA Pediatr.* **2020**. [CrossRef]
127. Bates, G.; Van Hout, M.C.; Teck, J.T.W.; McVeigh, J. Treatments for people who use anabolic androgenic steroids: A scoping review. *Harm Reduct. J.* **2019**, *16*, 75. [CrossRef] [PubMed]
128. Sessa, F.; Salerno, M.; Bertozzi, G.; Cipolloni, L.; Messina, G.; Aromatario, M.; Polo, L.; Turillazzi, E.; Pomara, C. miRNAs as novel biomarkers of chronic kidney injury in anabolic-androgenic steroid users: An experimental study. *Front. Pharmacol.* **2020**, *11*, 1454. [CrossRef] [PubMed]
129. Torrisi, M.; Pennisi, G.; Russo, I.; Amico, F.; Esposito, M.; Liberto, A.; Cocimano, G.; Salerno, M.; Li Rosi, G.; Di Nunno, N.; et al. Sudden Cardiac Death in Anabolic-Androgenic Steroid Users: A Literature Review. *Medicina* **2020**, *56*, 587. [CrossRef] [PubMed]
130. Mottram, D.R.; Alan, J.G. Anabolic steroids. *Best Pract. Res. Clin. Endocrinol. Metab.* **2000**, *14*, 55–69. [CrossRef]
131. Albano, G.D.; Sessa, F.; Messina, A.; Monda, V.; Bertozzi, G.; Maglietta, F.; Giugliano, P.; Vacchiano, G.; Marsala, G.; Salerno, M. AAS and organs damage: A focus on Nandrolone effects. *Acta Med. Mediterr.* **2017**, *6*, 939–946. [CrossRef]
132. Sessa, F.; Maglietta, F.; Bertozzi, G.; Salerno, M.; Di Mizio, G.; Messina, G.; Montana, A.; Ricci, P.; Pomara, C. Human brain injury and mirnas: An experimental study. *Int. J. Mol. Sci.* **2019**, *20*, 1546. [CrossRef]

133. Monda, V.; Salerno, M.; Sessa, F.; Bernardini, R.; Valenzano, A.; Marsala, G.; Zammit, C.; Avola, R.; Carotenuto, M.; Messina, G.; et al. Functional changes of orexinergic reaction to psychoactive substances. *Mol. Neurobiol.* **2018**, *55*, 6362–6368. [CrossRef]
134. Sessa, F.; Franco, S.; Picciocchi, E.; Geraci, D.; Chisari, M.G.; Marsala, G.; Polito, A.N.; Sorrentino, M.; Tripi, G.; Salerno, M.; et al. Addictions substance free during lifespan. *Acta Med. Mediter.* **2018**, *34*, 2081–2087.
135. Monda, V.; Salerno, M.; Fiorenzo, M.; Villano, I.; Viggiano, A.; Sessa, F.; Triggiani, A.I.; Cibelli, G.; Valenzano, A.; Marsala, G.; et al. Role of sex hormones in the control of vegetative and metabolic functions of middle-aged women. *Front. Physiol.* **2017**, *8*, 773. [CrossRef] [PubMed]
136. Pomara, C.; Neri, M.; Bello, S.; Fiore, C.; Riezzo, I.; Turillazzi, E. Neurotoxicity by synthetic androgen steroids: Oxidative stress, apoptosis, and neuropathology: A review. *Curr. Neuropharmacol.* **2015**, *13*, 132–145. [CrossRef] [PubMed]
137. Pomara, C.; D'Errico, S.; Riezzo, I.; De Cillis, G.P.; Fineschi, V. Sudden cardiac death in a child affected by Prader-Willi syndrome. *Int. J. Legal Med.* **2005**, *119*, 153–157. [CrossRef] [PubMed]
138. Fineschi, V.; Neri, M.; Di Donato, S.; Pomara, C.; Riezzo, I.; Turillazzi, E. An immunohistochemical study in a fatality due to ovarian hyperstimulation syndrome. *Int. J. Legal Med.* **2006**, *120*, 293–299. [CrossRef] [PubMed]
139. Neri, M.; Riezzo, I.; Pomara, C.; Schiavone, S.; Turillazzi, E. Oxidative-nitrosative stress and myocardial dysfunctions in sepsis: Evidence from the literature and postmortem observations. *Mediat. Inflamm.* **2016**, *2016*, 3423450. [CrossRef]

Article

Changes and Variations in Death Due to Senility in Japan

Masaki Bando [1,2,*], Nobuyuki Miyatake [1], Hiroaki Kataoka [3], Hiroshi Kinoshita [4], Naoko Tanaka [4], Hiromi Suzuki [1] and Akihiko Katayama [5]

1 Department of Hygiene, Faculty of Medicine, Kagawa University, Miki-cho, Kagawa 761-0793, Japan; miyarin@med.kagawa-u.ac.jp (N.M.); tanzuki@med.kagawa-u.ac.jp (H.S.)

2 Department of Physical Therapy, Kenshokai College of Health and Welfare, Tokushima 760-0093, Japan

3 Department of Physical Therapy, Faculty of Health Sciences, Okayama Healthcare Professional University, Okayama 700-0913, Japan; h.kataoka59@gmail.com

4 Department of Forensic Medicine, Faculty of Medicine, Kagawa University, Miki-cho, Kagawa 761-0793, Japan; kinochin@med.kagawa-u.ac.jp (H.K.); ntanaka@med.kagawa-u.ac.jp (N.T.)

5 Faculty of Social Studies, Shikokugakuin University, Zentsuji City 765-0013, Japan; kata@med.kagawa-u.ac.jp

* Correspondence: s19d736@stu.kagawa-u.ac.jp; Tel.: +81-86-891-2465

Received: 2 October 2020; Accepted: 27 October 2020; Published: 30 October 2020

Abstract: Objective: The proportion of elderly individuals (≥65 years old) in Japan has markedly increased. However, the definition of senility in Japan is controversial. The aim of the present study was to investigate changes and variations in the number of deaths due to senility in Japan. Methods: Information on the number of deaths due to senility between 1995 and 2018 as well as other major causes of death was obtained from the Statistics Bureau of Japan official website. Changes and variations in the number of deaths due to senility were compared with other major causes of death in Japan. The relationships between the number of deaths due to senility and socioeconomic factors were also examined in an ecological study. Results: The number of deaths due to senility was 35.7 ± 23.2/one hundred thousand people/year during the observation period and has continued to increase. A change point was identified in 2004 by a Jointpoint regression analysis. Variations in the number of deaths due to senility, which were evaluated by a coefficient of variation, were significantly greater than those due to other major causes of death, i.e., malignant neoplasm, heart diseases, cerebrovascular diseases, and pneumonia. The number of elderly individuals (≥65 years old) (%) and medical bills per elderly subject (≥75 years old) correlated with the number of deaths due to senility. Conclusion: The number of deaths due to senility has been increasing, particularly since 2004. However, variations in the number of deaths due to senility were observed among all prefectures in Japan.

Keywords: senility; Jointpoint analysis; coefficient of variation

1. Introduction

The proportion of elderly individuals (≥65 years old) in the total population of Japan has markedly increased, with 28.1% being older than 65 years in 2018 [1]. Among 1,362,482 deaths recorded in 2018 in Japan, the five major causes of death were malignant neoplasm (27.4%), heart diseases (15.3%), senility (8.0%), cerebrovascular diseases (7.9%), and pneumonia (6.9%) [2]. In turn, although older-age mortality will increase as the population ages, deaths will start to level off as the population shrinks in Japan, which is part of the demographic transition [3]. The number of deaths is expected to increase for some time in Japan [1]. Therefore, appropriate strategies for maintaining the health of the elderly are urgently needed.

Senility was the third major cause of death in Japan in 2018. Senility is only in the cause of natural death in the elderly when there is no other cause of death, and the death due to dementia is also defined apart from senility from the view of forensic medicine in the Ministry of Health, Labor and Welfare, Japan [4]. However, the definition of senility in Japan is controversial [5]. Esaki et al. [6] investigated the cause of death in subjects older than 100 years who died based on pathological anatomy, and identified relevant causes other than senile death in every case. Hawley [7] also reported the lack of senile deaths using a pathological autopsy. Gessert et al. [8] examined causes of death from the death certificates of 26,415 individuals aged 70 years and older, and suggested that death certificates need to be modified to facilitate the direct acknowledgment of age-related frailty as a contributing cause of death. In Japan, death due to senility is diagnosed only in the case of death due to natural causes, where there is no other cause of death in the elderly [4]. A study compared physicians with and without any experience of diagnosing death due to senility and demonstrated that there was no difference between the groups in terms of the average years of experience as a physician and years of experience in providing home care [9]. This suggests that rather than the level of experience, the beliefs and perspectives of the physicians engaged in terminal medical care have an impact on the number of death due to senility. Although it is one of the major causes of death in Japan, limited information is currently available on senility [10–13], particularly in recent years.

Therefore, in this study, we aimed to investigate the changes in the number of deaths due to senility, and to compare variations in the number of deaths due to senility with other major causes of death, i.e., malignant neoplasm, heart diseases, cerebrovascular diseases, and pneumonia in Japan.

2. Methods

2.1. Number of Deaths

Information on the number (/one hundred thousand people/year) of deaths due to the 5 major causes, i.e., malignant neoplasm, heart diseases, cerebrovascular diseases, pneumonia, and senility, between 1995 and 2018 (24 years) in 47 prefectures in Japan was obtained from the Statistics Bureau of Japan official website [14] (Table 1, Figure 1).

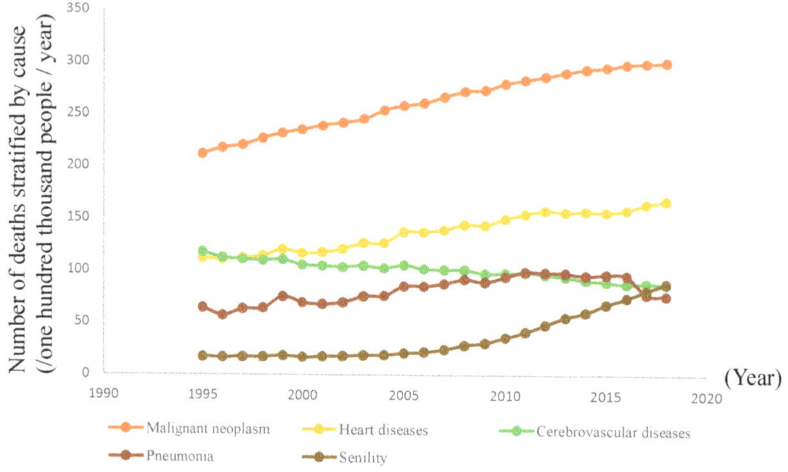

Figure 1. Changes in the number of deaths stratified by cause among all 47 prefectures in Japan.

Table 1. Number of deaths stratified by cause among all 47 prefectures in Japan.

	Mean ± SD			Minimum	Maximum
Number of Years		**24**			
Malignant neoplasm	261.8	±	28.9	211.6	300.7
Heart diseases	137.6	±	19.0	110.8	167.6
Cerebrovascular diseases	100.8	±	8.5	87.1	117.9
Pneumonia	81.2	±	13.3	56.9	98.9
Senility	35.7	±	23.2	16.7	88.2

Per 100,000 in 47 prefectures in every year for 24 years. SD: standard deviation.

2.2. Socioeconomic Factors

We also obtained information on socioeconomic factors, such as the number of elderly individuals (≥65 years old) (%), the number of single-person households (%), household income (Japanese yen), and medical bills per elderly subject (≥75 years old) (Japanese yen), which are considered to affect senility, from an official Japanese governmental website [15]. All parameters, except for the number of single-person households (%), were reported in 2016. The number of single-person households (%) was reported in 2015, but was not surveyed in 2016.

2.3. Ethics

The number of deaths due to major causes and socioeconomic factors were obtained from an official website. This study was approved by the ethical committee of Shikoku Gakuin University, Zentsuji city, Kagawa prefecture, Japan (approval number: 2020002, approval date: 8 September 2020).

2.4. Statistical Analysis

The number of deaths due to major causes in Japan (per 100,000 in 47 prefectures in every year) between 1995 and 2018 (24 years) were expressed as a mean ± standard deviation (SD). A trend analysis of the number of deaths due to senility was conducted by the Jointpoint regression program, 4.8.0.1 (National Cancer Institute) [16,17]. To compare variations in the number of deaths due to senility with other major causes of death in Japan, we calculated the coefficient of variation (CV: standard deviation/mean) of 5 major causes of death each year between 1995 and 2018 in Japan, and compared CV (among 47 prefectures for 24 years) using the Kruskal-Wallis test and Steel test [18]. We also examined the relationship between the number of deaths due to senility and socioeconomic factors in Japan using simple and multiple regression analyses, where $p < 0.05$ was significant. Correlation coefficient and partial correlation coefficient were calculated. Statistical analyses were performed using JMP Pro version 15 (SAS Institute Inc., Cary, NC, USA).

3. Results

Table 1 and Figure 1 show the number of deaths due to the five major causes between 1995 and 2018 in Japan. The total number of deaths due to malignant neoplasm was the highest among the five major causes, followed by heart diseases, cerebrovascular diseases, pneumonia, and senility. However, the number of deaths due to senility increased in recent years (Figure 1).

We investigated changes in the number of deaths due to senility by a Jointpoint regression analysis, which revealed a change point (2004) that was followed by marked increases (Figure 2).

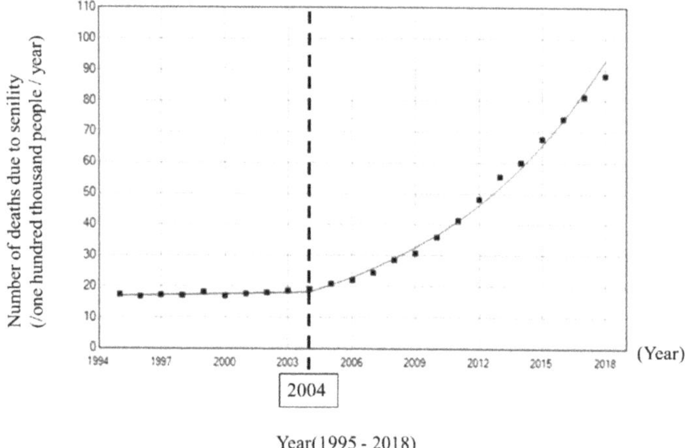

Figure 2. Jointpoint analysis of the number of deaths due to senility.

We then examined variations in the number of deaths due to the five major causes in Japan using CV (Table 2). The CV of senility was significantly higher than those of the other major causes during the time period analyzed (1995–2018). In addition, a change point (2004) was identified by a Jointpoint regression analysis. Therefore, we analyzed two periods (1995–2004 and 2005–2018), and found that the CV of senility was also higher than those of other causes in 1995–2004 and 2005–2018.

To identify the factors affecting the number of deaths due to senility, we examined socioeconomic factors, such as the number of elderly individuals (≥65 years old) (%), the number of single-person households (%), household income (Japanese yen), and medical bills per elderly subject (≥75 years old) (Japanese yen) in Japan, which are considered to be clinically important in 2016 (Table 3) in total, for men and women. A simple regression analysis showed that the number of elderly individuals (≥65 years old) (%), the number of single-person households (%), and medical bills per elderly subject (≥75 years old) (Japanese yen) correlated with the number of deaths due to senility in both sexes. Partial correlation coefficient between the number of deaths due to senility and the number of elderly individuals (≥65 years old) (%) was 0.696 ($p < 0.001$), and between the number of deaths due to senility and the medical bills per elderly subject (≥75 years old) (Japanese yen) was −0.512 ($p = 0.001$). In addition, we further analyzed this by using the number of deaths due to senility in 2017 and 2018. Correlation coefficient between the number of deaths due to senility and the number of elderly individuals (≥65 years old) (%) was 0.642 ($p < 0.001$) in 2017 and 0.657 ($p < 0.001$) in 2018, and correlation coefficient between the number of deaths due to senility and medical bills per elderly subject (≥75 years old) (Japanese yen) was −0.453 ($p = 0.001$) in 2017 and −0.434 ($p = 0.002$) in 2018.

In a multiple regression analysis, we used the number of deaths due to senility as a dependent variable and the number of elderly individuals (≥65 years old) (%), the number of single-person households (%), and medical bills per elderly subject (≥75 years old) (Japanese yen) as independent variables (Table 4). We found that the number of elderly individuals (≥65 years old) (%) and medical bills per elderly subject (≥75 years old) (Japanese yen) were important factors affecting the number of deaths due to senility in both sexes.

Table 2. Comparison of the coefficient of variation among deaths stratified by cause in all 47 prefectures in Japan.

	All		1994~2004 (Trend 1)		2005~2018 (Trend 2)	
	Mean ± SD	*p*	Mean ± SD	*p*	Mean ± SD	*p*
Malignant neoplasm	0.119 ± 0.003	**<0.001**	0.119 ± 0.005	**<0.001**	0.119 ± 0.002	**<0.001**
Heart diseases	0.162 ± 0.012	**<0.001**	0.150 ± 0.004	**<0.001**	0.170 ± 0.002	**<0.001**
Cerebrovascular diseases	0.219 ± 0.011	**<0.001**	0.208 ± 0.004	**<0.001**	0.226 ± 0.002	**<0.001**
Pneumonia	0.193 ± 0.016	**<0.001**	0.182 ± 0.009	**<0.001**	0.201 ± 0.004	**<0.001**
Senility	0.312 ± 0.038		0.349 ± 0.019		0.285 ± 0.006	

Bold values: $p < 0.05$ vs Senility as control group by the Steel test. SD: standard deviation.

Table 3. Relationships between the number of deaths due to senility and socioeconomic factors.

	All		Men		Women	
	r	*p*	r	*p*	r	*p*
The number of elderly individuals (≧65 years old) (%)	0.652	**<0.001**	0.539	**<0.001**	0.641	**<0.001**
The number of single-person households (%)	−0.460	**0.001**	−0.435	**0.002**	−0.455	**0.001**
Household income (Japanese yen)	−0.016	0.917	−0.001	0.995	−0.003	0.982
Medical bills per elderly subject (≧75 years old) (Japanese yen)	−0.439	**0.002**	−0.518	**<0.001**	−0.439	**0.002**

Bold values: $p < 0.05$ by a simple correlation analysis. r: correlation coefficient.

Table 4. Relationships between the number of deaths due to senility and socioeconomic factors by a multiple regression analysis.

All	B	95% CI		Standardized β	p	VIF
Constant	30.662	−34.577	95.901	0.000	0.349	
The number of elderly individuals (≧65 years old) (%)	4.961	3.387	6.534	0.641	**<0.001**	1.169
The number of single-person households (%)	−0.247	−1.475	0.980	−0.047	0.687	1.515
Medical bills per elderly subject (≧75 years old) (Japanese yen)	−0.0001	−0.0001	−4.260	−0.424	**<0.001**	1.352
$R^2 = 0.63, p < 0.001$						
Men						
Constant	30.540	−9.929	71.009	0.000	0.135	
The number of elderly individuals (≧65 years old) (%)	2.414	1.438	3.390	0.542	**<0.001**	1.169
The number of single-person households (%)	−0.024	−0.785	0.737	−0.008	0.950	1.515
Medical bills per elderly subject (≧75 years old) (Japanese yen)	−0.0001	−0.0001	−0.00003	−0.520	**<0.001**	1.352
$R^2 = 0.56, p < 0.001$						
Women						
Constant	52.772	−35.718	141.263	0.000	0.236	
The number of elderly individuals (≧65 years old) (%)	6.502	4.367	8.636	0.631	**<0.001**	1.169
The number of single-person households (%)	−0.315	−1.980	1.350	−0.045	0.705	1.515
Medical bills per elderly subject (≧75 years old) (Japanese yen)	−0.0001	−0.0002	−0.0001	−0.425	**<0.001**	1.352
$R^2 = 0.62, p < 0.001$						

Bold values: $p < 0.05$. 95% CI: 95% Confidence Interval. VIF: Variance Inflation Factor.

4. Discussion

In the present study, we examined changes and variations in the number of deaths due to senility in Japan. After 2004, the number of deaths due to senility significantly increased. The CV of the number of deaths due to senility was the highest among the five major causes of death in Japan.

The number of deaths due to senility has decreased since the 1950s [14]. Death due to senility was traditionally pronounced when there was no specific cause of death in the elderly, and was considered to be affected by the level of medical care. After the 1950s, improvements in medical treatment reduced the number of deaths due to senility. In the present study, the Jointpoint regression analysis revealed that the number of deaths due to senility increased after 2004. The number of elderly individuals and their mortality rate have both been increasing in Japan [2]. In the present study, simple and multiple regression analyses identified the number of elderly individuals as an important factor affecting the number of deaths due to senility. Collectively, the present results and previous findings indicate that the number of deaths due to senility will continue to increase in Japan.

As previous studies showed that the deaths due to senility is controversial [5], in the present study, the CV of senility was the highest among the five major causes, indicating variations in the definition of death due to senility. However, the CV of senility in 2005–2018 was lower than that in 1995–2004. The fact that the CV decreases in 2005–2018 may be important, and it shows that there is less variance in the more current data. Progression in diagnosis and better cause of death other than senility might affect these results.

In the present study, simple and multiple regression analyses showed that medical bills per elderly subject was one of the important factors affecting the number of deaths due to senility. Hasegawa et al. [19] reported that the mortality rate in hospitals and number of hospital beds per population were negatively associated with the number of deaths due to senility. Medical resources, including medical bills per elderly subject, the mortality rate in hospitals, and the number of hospital beds per population also appear to be associated with the number of deaths due to senility.

There were a number of potential limitations that need to be addressed. First, this was an ecological study and individual data were not obtained. Second, spurious correlation as anything that relates to an aging population may correlate with old-age mortality. Third, the socioeconomic factors examined in the present study may not have been sufficient. Other factors may affect the number of deaths due to senility. As Japan becomes a much older population, and as public health measures improve, we will see people dying at much older ages. This is a success story. If we cure cancer, heart disease and all other major causes of mortality, there will be an increase in mortality due to senility. Fourth, there was a limited number of years and the danger of extrapolating observations from data from a limited number of years. Validation studies are needed to confirm and extend our findings.

In conclusion, the number of deaths due to senility has increased and variations in senility were noted among the 47 prefectures in aging Japan. The definition of senility needs to be constant and detailed studies by using individual data are urgently required in the future.

Author Contributions: Conceptualization, M.B., and N.M.; investigation, M.B.; methodology, M.B., N.M., H.K. (Hiroaki Kataoka), H.K. (Hiroshi Kinoshita), N.T., H.S., and A.K.; formal analysis, M.B., N.M., and H.K. (Hiroaki Kataoka); Writing—Original draft preparation, M.B. and N.M.; Writing—Review and editing, M.B., N.M., H.K. (Hiroaki Kataoka), H.K. (Hiroshi Kinoshita), N.T., H.S., and A.K. All authors have read and agreed to the published version of the manuscript.

Funding: This research received no external funding.

Conflicts of Interest: The authors have no conflict of interest to declare.

References

1. Cabinet Office Official Website. Annual Report on the Ageing Society Summary 2019. Available online: https://www8.cao.go.jp/kourei/english/annualreport/2019/pdf/2019.pdf (accessed on 11 September 2020).
2. Ministry of Health, Labour and Welfare Official Website. Vital Statistics of Japan 2018. Available online: https://www.mhlw.go.jp/toukei/saikin/hw/jinkou/kakutei18/dl/00_all.pdf (accessed on 11 September 2020).
3. Takahashi, S.; Kaneko, A.; Ishikawa, A.; Ikenoue, M.; Mita, F. Population projections for Japan: Methods, assumptions and results. *Rev. Popul. Soc. Policy* **1999**, *8*, 75–115.
4. Ministry of Health, Labour and Welfare Official Website. Manual to Fill in a Death Certificate. Available online: https://www.mhlw.go.jp/toukei/manual/dl/manual_r02.pdf (accessed on 11 September 2020).
5. Traphagan, J.W. Localizing senility: Illness and agency among older Japanese. *J. Cross Cult. Gerontol.* **1998**, *13*, 81–98. [CrossRef] [PubMed]
6. Esaki, Y.; Sawabe, M.; Arai, T.; Matsushita, S.; Takubo, K. Pathologic evaluation of the main cause of death in Japanese centenarians. *Jpn. J. Geriat.* **1999**, *36*, 116–121. [CrossRef] [PubMed]
7. Hawley, C.L. Is it ever enough to die of old age? *Age Ageing* **2003**, *32*, 484–486. [CrossRef] [PubMed]
8. Gessert, C.E.; Elliott, B.A.; Haller, I.V. Dying of old age: An examination of death certificates of Minnesota centenarians. *J. Am. Geriatr. Soc.* **2002**, *50*, 1561–1565. [CrossRef] [PubMed]
9. Imanaga, T.; Toyama, T. Survey on the diagnosis of senility as the cause of death in home medical care. *Off. J. Jpn. Prim. Care Assoc.* **2018**, *41*, 169–175. [CrossRef]
10. Traphagan, J.W. Interpreting senility: Cross-cultural perspectives. *Care Manag. J.* **2005**, *6*, 145–150. [CrossRef] [PubMed]
11. Toda, T. Proteome and proteomics for the research on protein alterations in aging. *Ann. N. Y. Acad. Sci.* **2001**, *928*, 71–78. [CrossRef] [PubMed]
12. Kayukawa, Y.; Kogawa, S.; Tadano, F.; Imai, M.; Hayakawa, T.; Ohta, T.; Nakagawa, T.; Shibayama, H. Sleep problems in the aged in relation to senility. *Psychiatry Clin. Neurosci.* **1998**, *52*, 190–192. [CrossRef] [PubMed]
13. Ueda, K.; Hasuo, K.; Ohmura, K.; Kiyohara, K.; Kawano, H.; Kato, I.; Shinkawa, A.; Iwamoto, H.; Nakayama, K.; Omae, T. Causes of death in the elderly and their changing pattern in Hisayama, a Japanese community. Results from a long-term and autopsy-based study. *J. Am. Geriatr. Soc.* **1990**, *38*, 1332–1338. [CrossRef] [PubMed]
14. Personal Site of Official Statistics of Japan. Explore Japanese Government Statistics. Available online: https://www.e-stat.go.jp/stat-search/files?page=1&layout=datalist&toukei=00450011&tstat=000001028897&cycle=7&tclass1=000001053058&tclass2=000001053061&tclass3=000001053065 (accessed on 11 September 2020).
15. Personal Site of Official Statistics of Japan. Explore Japanese Government Statistics. Available online: https://www.e-stat.go.jp/stat-search/files?page=1&layout=datalist&toukei=00200502&tstat=000001137289&cycle=0&year=20200&month=0&tclass1=000001137291 (accessed on 11 September 2020).
16. Yang, L.; Sakamoto, N.; Marui, E. A study of home deaths in Japan from 1951 to 2002. *BMC Palliat. Care* **2006**, *5*, 2. [CrossRef] [PubMed]
17. JACR Newsletter. Japanese Association of Cancer Registries. Available online: http://www.jacr.info/publicication/Pub/NL/NL46/NL_46-8P.pdf (accessed on 11 October 2020).
18. Bando, M.; Miyatake, N.; Kataoka, H.; Kinoshita, H.; Tanaka, N.; Suzuki, H.; Katayama, A. Relationship between air temperature parameters and the number of deaths stratified by cause in Gifu prefecture, Japan. *Healthcare* **2020**, *8*, 35. [CrossRef] [PubMed]
19. Hasegawa, T. Prefectural difference in percentage of death due to senility in Japan. *Bull. Soc. Med.* **2017**, *34*, 89–93.

MDPI
St. Alban-Anlage 66
4052 Basel
Switzerland
Tel. +41 61 683 77 34
Fax +41 61 302 89 18
www.mdpi.com

Healthcare Editorial Office
E-mail: healthcare@mdpi.com
www.mdpi.com/journal/healthcare

www.ingramcontent.com/pod-product-compliance
Lightning Source LLC
LaVergne TN
LVHW070649170726
843515LV00004B/360
* 9 7 8 3 0 3 6 5 4 1 9 4 5 *